Philosophy of science

Fundamentals of Philosophy
Series editor: John Shand

This series presents an up-to-date set of engrossing, accurate, and lively introductions to all the core areas of philosophy. Each volume is written by an enthusiastic and knowledgeable teacher of the area in question. Care has been taken to produce works that while evenhanded are not mere bland expositions, and as such are original pieces of philosophy in their own right. The reader should not only be well informed by the series, but also experience the intellectual excitement of being engaged in philosophical debate itself. The volumes serve as an essential basis for the undergraduate courses to which they relate, as well as being accessible and absorbing for the general reader. Together they comprise an indispensable library of living philosophy.

Published:
Piers Benn
Ethics

Alexander Bird
Philosophy of Science

Colin Lyas
Aesthetics

Alexander Miller
Philosophy of Language

Forthcoming:
Stephen Burwood, Paul Gilbert, Kathleen Lennon
Philosophy of Mind

Richard Francks
Modern Philosophy

Dudley Knowles
Political Philosophy

Harry Lesser
Ancient Philosophy

Philosophy of science

Alexander Bird

University of Edinburgh

First published in 1998 by UCL Press

Reprinted 2000

UCL Press Limited
11 New Fetter Lane
London EC4P 4EE
UK

UCL Press is an imprint of the Taylor & Francis Group

British Library Cataloguing in Publication Data
A catalogue record for this book is available
from the British Library.

ISBNs:
1-85728-681-2 HB
1-85728-504-2 PB

Typeset in Century Schoolbook and Futura.
Printed and bound by
T.J. International Ltd., Padstow, Cornwall

For Michael Bird

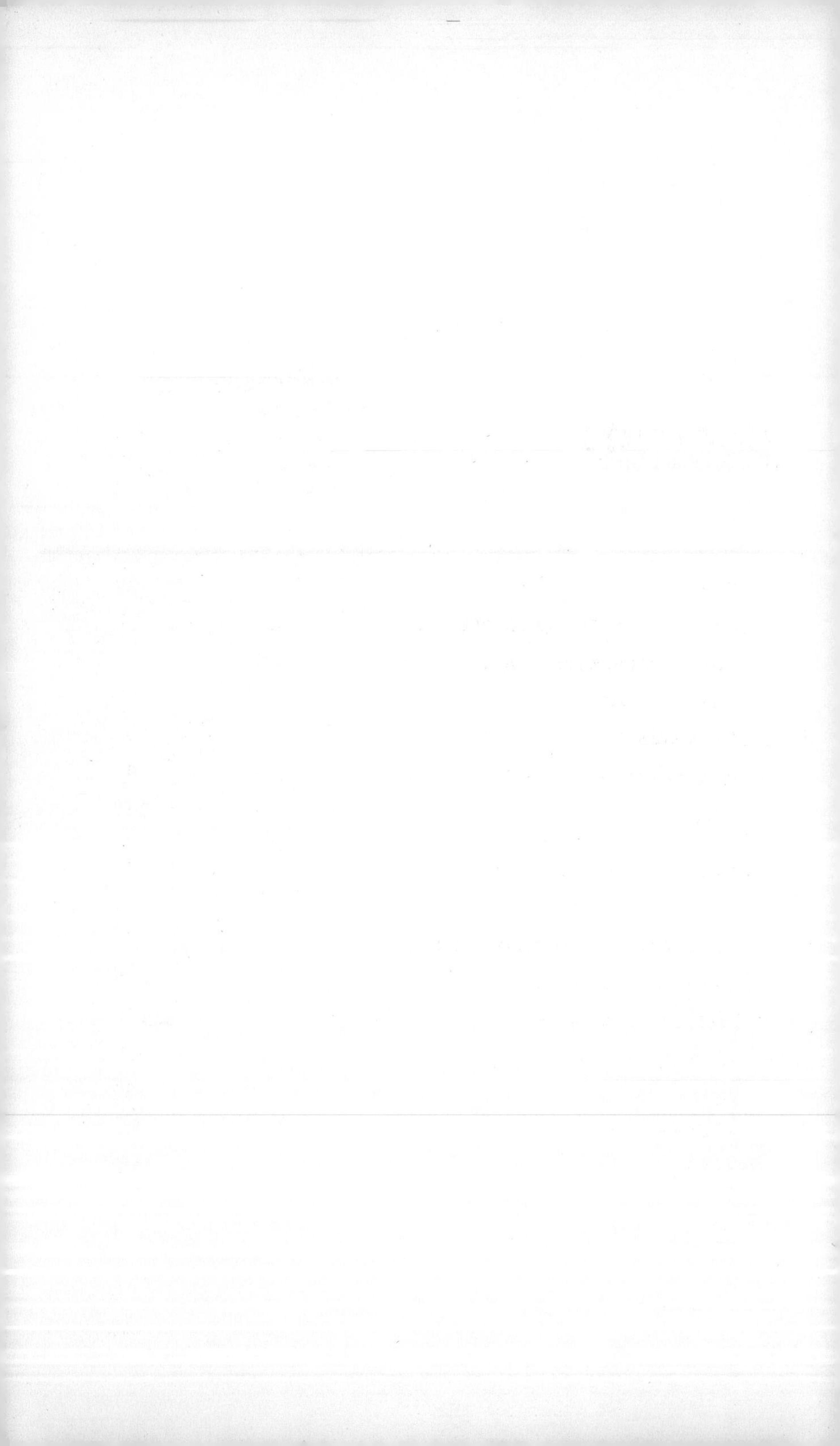

Contents

Preface

This book aims to be an introduction to the philosophy of science. While I expect students of philosophy to form the majority of its readers, I hope that it will also find favour among students of science subjects and indeed anyone else who is interested in conceptual questions surrounding the nature of the world and our scientific investigation of it. Conceptions of what the philosophy of science is have changed markedly over the years. For a long time positivist concerns were at the fore, for instance establishing the observational given from which scientific reasoning starts (the problem of the empirical basis), or constructing an inductivist logic or theory of confirmation which would take us from the empirical basis to generalized scientific knowledge. The positivist and related programmes did not succeed. There ensued a reaction in which this failure was taken to demonstrate the hopelessness of any objective conception of scientific knowledge and progress. Instead the appropriate task for philosophers of science was to provide support for more historical and indeed sociological approaches to science. This road was fruitful in many ways. But, especially by associating itself with relativism about knowledge and progress, it divorced itself from work being done in the mainstream of philosophy, especially in metaphysics and epistemology.

My impression is that the philosophy of science is swinging back the other way. But not to where it began. Philosophers of science are now much more historically sensitive. They are not likely to put inductive logics or the problem of the empirical basis at the centre of their concerns. Nonetheless, their interests have moved closer to those of philosophers in other fields. This book is intended both to contribute to as well as to describe that shift. Thus the first part of the book is primarily metaphysical, seeking to give a philosophical account of the most general sorts of thing scientists aim to represent – laws of nature, explanations, natural kinds. The second half moves into epistemological territory and sets out to show how an externalist, moderately naturalized epistemology can show how scientific knowledge is possible. This leads to a view about the scientific method (that there is no such unique thing) and progress (it is possible) into which I weave a discussion of the historical approach.

The ideas of many philosophers, colleagues, and friends have gone into making this book possible. The suggestions for further reading, the notes, and the bibliography give some partial but incomplete indication of my debt to other philosophers working in this area. What I owe to friends and acquaintances with whom I have exchanged thoughts is even less well indicated. Nonetheless, I would like to thank in particular those who have read parts of earlier drafts and thereby enabled me to complete this work: Helen Beebee, Rebecca Bryant, Timothy Chan, Penn Dodson, Richard Foggo, Will Hubbard, Scott Jacobs, Mary MacLeod, Hugh Mellor, Lucasta Miller, Peter Milne, Jim Moor, John Pellettieri, John Shand, Walter Sinnott-Armstrong, Lucy Wayman, Denis Walsh, Timothy Williamson, and Ben Young.

Introduction

The nature of science

Our starting point is the question *What is science?* In the first part of this chapter we will see what an American judge had to say when the question came before his court. He had to rule on whether what is called "creation science" really is science. This is useful, first because it shows that the question is a live and important one, and secondly because the answer is revealing. Although the judge did not give a complete and systematic answer (as a philosopher may have attempted to do), he did identify several of the key issues. These are the subject matter for much of the rest of this book. The judge was interested in questions like *What is science about?* and *What is a scientific theory?* He did not, however, ask *How do we know when a theory is true?*, mainly because he was not concerned with the truth of creation science, but with whether creationism is science or not. For a theory can be false yet still be scientific (e.g. Priestley's phlogiston theory of combustion), and a claim can be true without being scientific (e.g. the claim that Brazil won the 1994 World Cup). Nonetheless, the nature of scientific knowledge is clearly important and this chapter ends with two puzzles concerning it.

What is science?

First, the background. In recent years an interesting phenomenon has appeared in American political, religious, and educational circles. For instance, in 1995 and 1996 new laws were proposed in the legislatures of five American states that required schools to give equal attention to evolution and to creation science in state school science classes. Creation science originates with a literal reading of the Bible, in particular the book of Genesis. It claims, however, not to be religious in content; rather, it says, the relevant propositions, for instance that the universe, the solar system, and life were suddenly created, that currently living animals and plants have existed since creation, that this creation was relatively recent (i.e. some thousands of years ago), that man and apes have a separate ancestry, and that the geological features of the Earth are to be explained by large-scale sudden catastrophes (e.g. a flood) are all scientific hypotheses for which there is strong scientific evidence.[1]

Why was the Genesis story turned into science in this way? For centuries scientists have regarded the universe and its laws as the manifestations of the will and nature of its Creator. And it is natural enough to seek in it signs of those events that the Bible records as His work. More recently, however, creation science has gained a political purpose. The constitution of the USA provides for the separation of church and state, and for this reason religious doctrine may not be taught in state schools. Nonetheless, for many years some states have prohibited the teaching of those scientific theories that would be antithetical to the prevailing fundamentalist version of Christianity. It is for this reason that the famous "monkey" trial of John Scopes took place, in Dayton, Tennessee, in 1925. Scopes was accused, and found guilty, of violating a Tennessee law forbidding the teaching of evolution. Such laws no longer exist. (It may have been the fear of communism that brought the teaching of evolution to those schools where it had been ignored. The launch of Sputnik in 1957, the first artificial satellite, raised the spectre of the USSR stealing a march on the USA in science and technology, and thereby led to a revival in science education, although Tennessee's Butler Act forbidding the teaching of evolution remained on the statute books until 1967.)

Since the end of the Cold War, Christian fundamentalists, allied with elements of the political right wing (a coalition known as the Religious Right), have sought to regain their position in American education. But the educational and political climate has changed so much that there could be no return to an outright ban on evolutionist teaching. The next best thing for the Religious Right would be the introduction, on equal terms, of the creationist alternative into the syllabus. Even so, as creationism is a religious doctrine, teaching it in state schools would clearly violate the First Amendment of the Constitution regarding separation of church and state. So, in order to get creationism introduced into schools, and, better, into the same classes in which evolution is taught, supporters of creationism have sought to pass it off as a purely scientific doctrine devoid of religious content. In their words "*the creation model is at least as scientific as the evolution model, and is at least as nonreligious as the evolution model.*" [2]

Indeed, the state of Arkansas passed a law that required equal treatment of evolution and creationism. Scientists and churchmen immediately challenged the constitutionality of the new law in the courts. They argued that, despite the rhetoric and the appearance of scientific respectability, creation science simply is not science but religion dressed up as science. What is at issue is not science versus religion (many religious people reject creationism), or the question of whether creationism is true or even reasonable. The debate focuses on the claims of creation science to be *scientific*. So the court had to ask: what is science? When is a claim scientific? How do we distinguish science from non-science or pseudo-science? These are all philosophical questions, and the court had the benefit of the testimony of philosophers as well as scientists in its deliberations. The opinion of the judge, William R. Overton, is interesting in this respect, because he summed up the criteria he used in deciding whether creationism is scientific.[3] He said that a scientific theory has the following features:

(a) It is guided by natural law.
(b) It has to be explanatory by reference to natural law.
(c) It is testable against the empirical world.
(d) Its conclusions are tentative, i.e. are not necessarily the final word.
(e) It is falsifiable.

What did Overton mean by these criteria? The creationist claims about the origins of the world, in order to be science rather than mere assertion, must be claims (a) about what the laws of nature are, or (b) about facts which themselves have explanations in terms of natural law. (I will tend to use the term "law of nature" for what Overton calls "natural law"). What are laws of nature? This is a question for the next chapter, but roughly speaking they are the very general relations among things that explain their behaviour – the sort of things that we call the laws of physics, chemistry, and so on. Newton's laws of motion, Ohm's law of resistance, Hooke's law of elasticity, and so on are things we think or have thought are laws. (c) The claims must be made in such a way as to be testable, i.e. be sufficiently detailed as to have observable consequences or testable predictions that can be verified. (e) In particular, if these testable claims are found to be refuted by experience, then this should count as powerful evidence against the theory, which may amount to showing the theory to be false. Our belief in the theory should be proportional to the strength of the evidence in its favour (which means considering it in relation to its rivals). (d) Even if the evidence is very favourable, experience shows that sophisticated well-confirmed theories can be overturned by falsifying evidence found later or challenged by new competing theories that themselves are supported by the evidence in the way that the original theory is not. A scientific attitude requires being open to this possibility.

Let us look at some examples. The creationist says that the Earth's geological features are explained in terms of catastrophic events, primarily a large-scale flood. By contrast, standard geology explains them in terms of forces that work over extended periods of time – erosion, glaciation, plate tectonics. The latter can all be understood as phenomena that obey the laws of physics. For instance, tidal motion is a result of the Moon's gravitational pull, and so is a manifestation of the law of gravity. Even if a massive flood could explain the Earth's geology, why did the flood occur in the first place? While disastrous floods do still happen today, they are restricted in scope and can to some extent be explained by reference to local weather conditions such as hurricanes. But a global flood is a different matter. What sort of weather conditions could bring about such a calamity? If the whole Earth was covered in water, where did it all come from? And where did it go to thereafter? No laws or law-governed phenomena are invoked – or could be – to answer such questions. The reason why is clear. A flood, Noah's, is referred to in the book of Genesis. This flood was the work of divine intervention. While God's actions may indeed be the correct explanation of an event, such an explanation fails to be scientific by Judge Overton's criteria. God's will is not subject to natural law. And for precisely that reason this hypothesis is not testable either. God does not submit to experimentation.

Of course all explanations end somewhere, even orthodox ones. The difference here is that the orthodox scientist seeks to extend the chain of explanations as far as possible and is limited only by ignorance, while the creationist is happy to terminate the chain of explanations much earlier, in accordance with biblical constraints, and at that point to use supernatural explanations.

Another reason creationists have for adopting catastrophism is that they are committed to a young Earth – six to ten thousand years old – too young for the long-term effects of traditional geological processes. The nineteenth century physicist Lord Kelvin (whose name was given to the scale of temperature) made careful calculations concerning the rate of cooling of the Earth and Sun. These suggest that the solar system could not have been around long enough for life to have evolved. His arguments are still quoted by creationists. Kelvin's calculations were quite right – if one assumes no heat source in the Earth and that the Sun's energy comes from combustion. In both cases there are energy sources he ignored, viz. nuclear processes, fusion in the Sun, and radioactive decay in the Earth. Lord Kelvin can be forgiven for not knowing about something discovered only after he made his calculations; a contemporary writer cannot. Again the best explanation of the creationists' stance is not the evidence and arguments that they present, but a prior commitment to biblical historiography (from which the date of Adam's creation and so the creation of everything can be calculated).

Questions about the origins of the Earth exhibit this contrast of approach most clearly. Standard science has an explanation of the birth of the solar system that shows how gravitational forces within clouds of gas would condense the gases to stars and planets. The laws of nature we have discovered allow us to trace back the history of the universe to its earliest moments. Observation of the red-shift of galaxies allows us to formulate Hubble's law relating the speed of the recession of those galaxies to their distance. This in turn gives us some idea of the rate of expansion of the universe and hence its age. The creationists have nothing to say along such lines. The Earth and Sun and life were created simultaneously, several thousand years ago, as an act of God. How do we know this? Judge Overton quotes a leading creationist authority:

> ... it is ... quite impossible to determine anything about the Creation through a study of present processes, because present processes are not creative in character. If man wishes to know anything about Creation (the time of Creation, the duration of Creation, the order of Creation, the methods of Creation, or anything else) his sole source of true information is that of divine revelation. God was there when it happened. We were not there ... Therefore we are completely limited to what God has seen fit to tell us, and this information is in His written Word. This is our textbook on the science of Creation![4]

Such an approach to the facts of creation is scarcely "tentative". Not only is it the final word, it is the Word. If creation were not subject to natural laws, then no amount of scientific investigation could have any bearing on it. This aspect of creationism is independent of natural explanation. It has no need of being tentative. Faith may be strong or it may be weak, but it is not a hypothesis. As creationism is not a hypothesis that invokes natural law, it is not open to amendment or refutation in the face of experience. The above quotation makes it clear that, as creation is quite a different sort of happening from anything occurring today, no observation of the latter could bear on claims about the former. The point of remarks such as these is not to refute any belief in divine creation, but to examine the claim that such beliefs can be *scientific*. There may be non-scientific knowledge of and justification for beliefs in creation; but that is another question. It may be worth remarking that there are few, if any, people who have adopted creationism *without* being religious and *because* of the supposedly scientific arguments for it.

Creationists are fond not only of displaying the scientific

credentials of their arguments but also of pointing out that evolutionists depend upon faith as much as anyone. This refers to the fact that there are unsolved problems concerning evolution. In the face of such difficulties, the scientists' belief in evolution must be a matter of faith. While most of these unsolved anomalies are in fact spurious or have already been solved (e.g. Kelvin's objection), it must be admitted that there are others for which a satisfactory solution remains to be found. For instance, the processes whereby life emerged from the primaeval soup are poorly understood. While we know how amino acids, the building blocks of organic molecules, can be created in a primitive atmosphere, there is a problem with understanding the origin of proteins. The interaction of proteins is not merely chemical but also depends on their large-scale architecture. It is not clear how proteins could come into being and replicate. Nonetheless, a scientist's rational adherence to a theory does not require it to be flawless, let alone be without a problem. Scientists require a theory to be a good, preferably the best available, explanation of a range of important phenomena. This in turn depends on the theory's ability to generate and maintain strategies for solving problems.

So the sign of a successful theory is not the absence of problems but its ability to solve those problems which do arise. An instance of this is explained by Stephen Jay Gould in his famous essay *The panda's thumb*. The giant panda appears to have six fingers, one of which is an opposable thumb useful for gripping things such as bamboo stalks. This is very odd. Primates, but few other animals, have an opposable thumb. According to the story of primate evolution this is a development of one of five digits. Bears have the same five digits but have not developed any of them into a thumb, because their paws have evolved, like those of many animals, for the purpose of running. Where does the panda's thumb come from? Being a sixth finger it cannot have developed from one of the pre-existing five, as in the case of the human thumb. But the evolution of a completely new finger, though perhaps possible, seems unlikely. (Perhaps one might wonder if the panda did not evolve but was put on Earth properly designed for its purpose.) It turns out that the panda's thumb is not a true finger at all. There are no new bones involved. Rather an existing bone, part of the panda's wrist, has become elongated until it is able to function like an opposable thumb. Evolutionarily this is highly plausible. Gould's point is that nature does not show evidence of design but the adaptation for a new purpose of organs and limbs that first evolved with a different purpose (which he says is more like tinkering than design). My point, more broadly, is that the ability of evolutionary theory to cope with such oddities is the reason why scientists find it credible. As long as it continues to be successful in this way the appearance of new problems, far from being a sign of failure, is what provides fuel for growing knowledge. Of course it may be that the problems begin to pile up without the recompense of adequate solutions, and then we may be in a different ballpark. This is an interesting issue and we shall come back to it later in this book.

Judge Overton's reasons for rejecting creation "science" are useful to us not only in providing criteria for deciding what may or may not properly be called science, but also because they highlight many of the ideas and notions that we as philosophers of science must investigate. His analysis says that science should concern natural laws or explanations in terms of natural law. But what is a natural law and what counts as an explanation? While these concepts may have been clear enough for the judge's purposes, we shall see that once one looks at them at all closely all sorts of problems begin to arise.

A claim of the creationists which the court rejected was that living things come in distinct kinds. The judge and the witnesses against the State of Arkansas seemed to think that this concept is unscientific. Perhaps one can see why, given the creationists' use of it. In their view, mankind belongs to a different kind from other primates. Yet they also claim that all the different species of bats (over 800 of them forming a diverse order) belong to a single kind. This seems an odd and tendentious notion of kind. But surely there is some sense in saying that chemical elements make up distinct kinds? And what about different species? We shall see whether there is a need for a notion of *natural kind* and what that notion is.

Judge Overton says that a scientific theory should be both testable and falsifiable. Are these different or the same? Sir Karl Popper maintained that a theory could only be tested by attempting to falsify it. Success in passing such tests corroborates a theory. But much of the evidence for Darwin's hypotheses came from the fossil record. Creationists frequently point out the existence of gaps in the fossil record – the lack of fossils for intermediate evolutionary types. Darwin himself was unconcerned by this – the fossil record is just incomplete. And so the existence of gaps should not be taken as a refutation. But, on the other hand, the existence of expected fossils is positive evidence. So observations of fossils could confirm but not refute the Darwinian thesis. What then is falsifiability? And when is a theory confirmed by evidence?

When we were discussing laws, explanations, and natural kinds we were concerned with the subject matter, in the most general terms, of a scientific theory. Now we have come to questions like: How may we falsify or confirm a theory? When is a scientific belief justified? When does a scientific belief amount to knowledge? The most extraordinary thing about science is the depth and range of the knowledge it claims to be able to give us, and it is about this knowledge that the philosophical puzzles arise. Without belittling the wisdom of societies without modern science, one may still wonder at the achievement – or arrogance, if you prefer – of scientists who claim some understanding of what happened during the first few moments of the existence of the universe, of the laws that govern the motions of the stars and planets, of the forces and processes that gave birth to life on Earth, and of the structure, functioning, and origin of living organisms.

The story of why this knowledge came about when and where it did, is one for the historian and sociologist of science. Our job as philosophers is not to ask where this knowledge came from, but to look carefully at that knowledge itself and ask: What is it? Is scientific knowledge really as it first appears to us to be? This enquiry I will break down into two parts. The first asks, What is scientific knowledge knowledge of? The straightforward answer is: science is knowledge of why things are the way they are, of the kinds of thing there are to be found in nature, and of the laws governing these things. We shall see that things are not so simple when it comes to saying what a law of nature is, what an explanation is, and what a natural kind is. Indeed, we shall see that the idea that science provides knowledge of a reality independent of humans is itself intensely problematic. These issues will occupy us in Chapters 1 to 4.

Having considered what scientific knowledge is knowledge of, the second part of our enquiry goes on to ask what right we have to think of it as knowledge at all? Perhaps what we call knowledge is not knowledge at all – perhaps it just appears to be knowledge. What would make the difference? These questions come in the second part this book (Chs. 5–8), since the importance of our knowledge as well as our route to it depend upon what it is we are supposed to know about. Nonetheless, while investigating the concepts of law, kind, and explanation, we will need to bear in mind quite general problems that face anyone who claims to have scientific knowledge. Philosophy is best approached through its problems and puzzles, and the philosophy of science is no exception to this. So we shall start by introducing two puzzles about scientific knowledge (which I will call Hume's problem and Goodman's problem). They both concern *induction*, which is at the heart of scientific reasoning.

What is induction?

To set the stage then for these puzzles, I shall say something about what induction is. As the nature of induction is so puzzling, it would be foolish to attempt to define it. But some rough descriptions will give you the idea. We use "induction" to name the form of reasoning that distinguishes the natural sciences of chemistry, meteorology, and geology from mathematical subjects such as algebra, geometry, and set theory. Let us look then at how the way in which scientific knowledge is gained differs from the reasoning that leads to mathematical understanding. One distinction might be in the data used. For instance, while knowledge in the natural sciences depends upon data gained by observation, the practitioners of mathematical subjects do not need to look around them to see the way things actually are. The chemist conducts experiments, the geologist clambers up a rock face to observe an unusual rock stratum, and the meteorologist waits for data from a weather station, whereas the mathematician is content to sit at his or her desk, chew a pencil, and ruminate. The mathematician seems to have no data, relying instead on pure thought to conjure up ideas from thin air. Perhaps this is not quite right, as the mathematician may be working with a set of axioms or problems and proofs laid down by another mathematician, or indeed with problems originating in real life or in the natural sciences, for instance Euler's Königsberg bridge problem[6] or some new theory required by subatomic physics. These things are not quite like the data of geology or the experimental results of chemistry, but let us agree that they at least constitute a basis or starting point for reasoning that will lead, we hope, to knowledge.

What really is different is the next stage. Somehow or other the scientist or mathematician reaches a conclusion about the subject matter, and will seek to justify this conclusion. In the case of the scientist the conclusion will be a *theory*, and the data used to justify it are the evidence. The mathematician's conclusion is a *theorem* justified by reference to axioms or premises. It is the nature of the justification or reasoning that differentiates the two cases; the mathematician and the scientist use different kinds of argument to support their claims. The mathematician's justification will be a proof. A proof is a chain of reasoning each link of which proceeds by deductive logic. This fact lends certainty to the justification. If what we have really is a mathematical proof, then it is certain that the theorem is true so long as the premises are true. Consequently, a proof is final in that a theorem once established by proof cannot be undermined by additional data. No one is going to bring forward evidence contrary to the conclusion of Euclid's proof that there is no largest prime number.[7] This contrasts with the scientific case. For here the support lent by the data to the theory is not deductive. First, the strength of justification given by evidence may vary. It may be very strong, it may be quite good evidence, or it may be very weak. The strength of a deductive proof does not come by degree. If it is a proof, then the conclusion is established; if it is flawed, then the conclusion is not established at all. And, however strongly the evidence is shown to support a hypothesis, the logical possibility of the hypothesis being false cannot be ruled out. The great success of Newtonian mechanics amounted to a vast array of evidence in its favour, but this was not sufficient to rule out the possibility of its being superseded by a rival theory.

All these differences between mathematics and the natural sciences characterize the distinction between deductive and non-deductive reasoning. Not only mathematical reasoning is deductive. For example, the following argument is deductive:

All mammals suckle their young;
otters are mammals;
therefore
otters suckle their young.

The next is deductive too:

Duck-billed platypuses do not suckle their young;
platypuses are mammals;
therefore
not all mammals suckle their young.

The point about these deductive arguments, like mathematical ones, is that their assumptions or premises *entail* their conclusions. By "entail" we mean that should the premises be true then, as a matter of logic, the conclusion must also be true.

In science, inferences from data to generalizations or to predictions typically do not entail their conclusions; they do not carry the logical inevitability of deduction. Say a doctor observes that all their patients suffering from colitis also suffer from anaemia. They might hypothesize that they go together, so that everyone who has colitis is also anaemic. Further observation that all of a large variety of colitis patients have anaemia would be regarded as evidence in favour of this hypothesis. But, however many such patients we observe, we have not ruled out the logical possibility of a colitis patient without anaemia. Similarly, overwhelming evidence in favour of the theory that the extinction of the dinosaurs was caused by a meteorite impact does not logically rule out all alternative hypotheses. (The evidence may rule out all sensible alternative hypotheses, for instance that the cause was volcanic eruption. But some explanations cannot be ruled out. For instance, one creationist view is that the fossil record was specially created by God as if there had been dinosaurs. One could imagine similar hypotheses relating to the evidence for meteorite impact. Perhaps one could hypothesize that the dinosaurs were executed by criminal extraterrestrials, who carefully covered their tracks.)

These scientific, non-deductive arguments are often called *inductive*. We have to be a little careful here because the word "inductive" is used in at least two different ways. In its broadest sense it means just "non-deductive" and is thereby supposed to cover all the kinds of argument found in science (other than any deductive ones). So, the great historian and philosopher of science, Sir William Whewell spoke of the inductive sciences simply to contrast them with the deductive sciences of logic and the various branches of mathematics. "Induction" has also been used to name a more specific kind of scientific argument, i.e. one where we argue from several particular cases to the truth of a generalization covering them. We have just seen the case of a doctor who took the existence of many anaemic colitis sufferers to be evidence for the claim that all colitis sufferers have anaemia; this was inductive reasoning in this second, more specific sense. The hypothesis that meteorite impact caused the extinction of the dinosaurs can be justified by the evidence, but this justification is inductive only in the former, broader sense. We need to distinguish between these two senses: I am happy to keep using the word "induction" with the broader sense; the narrow sense I will indicate by the phrase "Humean induction", for reasons that will become clear shortly.

We saw earlier on that some forms of knowledge seem to be obtainable just by pure thinking while others require the making of observations and the gathering of data. Philosophers mark this distinction by the use of the terms *a priori* and *a posteriori*. *A priori* knowledge is knowledge that can be gained without the need for experience. (To be precise, *a priori* knowledge is knowledge that can be gained without any experience *beyond that required to acquire the relevant concepts*. This additional proviso is required because, without some experience, one may not be able to have the thought in question.) *A posteriori* knowledge is knowledge that is not *a priori*. So to have *a posteriori* knowledge experience is required. Pure mathematics and logic are usually taken to be examples of *a priori* knowledge, whereas most of chemistry and biology are *a posteriori*. Another concept used in this connection is that of an *empirical proposition*. Empirical propositions are those the truth or falsity of which can only be known *a posteriori*.

If we take a typical generalization of science – all mammals possess a muscular diaphragm – we can see that it is empirical.

Knowing what this proposition means is not sufficient for knowing whether it is true or false. To know that, we would have to go and examine at least some mammals. Then we would have to infer from our observations of a limited range of mammals that all, not just the ones we have seen, possess a muscular diaphragm. This is an inductive inference.

It is often said that *a priori* knowledge is certain, and by implication empirical propositions like those discussed cannot be known with certainty. "Certainty" is a slippery concept and is related to an equally slippery concept, "probability", which I will discuss in Chapter 6. One reason why *a priori* knowledge is thought to be certain in a way that *a posteriori* knowledge of empirical generalizations is not, is that the former is gained through deductive reasoning while the latter requires inductive reasoning. As we saw above, if the premises of a deductively valid argument are true, the conclusion must be true too, while the premises of an inductive argument might be true yet the conclusion still false. But we need to be careful here, for as we use the word "certain", we are certain of many empirical propositions: that we are mortal, that the water at 30°C in my bath tonight will be liquid, and that this book will not turn into a pigeon and fly away. If someone asks why we are certain we may appeal to our past experience. But past experience does not entail these claims about the future. Whatever the past has been like, it is logically possible that the future will be different. One might argue from some natural law. It is a law of nature that people will die and that water at 30°C is liquid. How are we certain about these laws of nature? Here again it is because we have reasoned on the basis of our experience that these are among the laws of nature. But this experience does not entail the existence of a law. For it may have been sheer chance that so many people have died, not a matter of law. So to reason that there is such a law and that we too are mortal is to reason inductively, not deductively. Yet it is still something of which we are certain, and in this sense induction can give us inductive certainty.

So far I have said nothing to impugn the integrity of inductive argument; all I have done is to draw some rough distinctions between deductive and inductive reasoning. The fact that deductively valid arguments entail their conclusions in a way that inductive arguments do not, may suggest that induction is inferior to deduction. This would be misleading. Consider the deductive inference: all colitis sufferers are anaemic, therefore this colitis sufferer is anaemic. You can see that the conclusion "this colitis sufferer is anaemic" says nothing more than is already stated in the premise "all colitis sufferers are anaemic". To say all colitis sufferers are anaemic is to say, among other things that this colitis sufferer is anaemic. So logical certainty goes along with saying nothing new. But in saying something about the future we will be adding to what we have already said about the past. That is what makes induction interesting and valuable. There would be no point in wanting induction to give us logical certainty, for this would be tantamount to wanting it to tell us nothing new about the future but only to restate what we already knew about the past.

Hume's problem

There is nonetheless a well-known difficulty associated with induction. The problem is this: Can these inductive arguments give us knowledge? This dilemma was first identified by the great Scottish historian and philosopher David Hume (1711–76) whose views on causation and induction, as we shall see, have had a very strong influence on the development of the philosophy of science. Although the dilemma is framed in terms of Humean induction, it is in fact perfectly general and raises problems for any form of induction. (Hume himself posed the problem in terms of causation.) It proceeds as follows. If an argument is going to give us knowledge of the conclusion, then it must justify our belief in the conclusion. If, because we have observed all colitis patients hitherto to have anaemia, we argue that all colitis sufferers are anaemic, then those observations along with the form of the inductive argument should justify the belief that all colitis sufferers are anaemic.

How does this justification come about? On the one hand, it cannot be the sort of justification that a deductive argument gives to its conclusion. If it were, the conclusion would be a logically necessary consequence of the premises. But, as we have seen, this is not the case; in inductive arguments it is logically possible for the conclusion to be false while the premises are true. The very fact that inductive arguments are not deductive means that the sort of justification they lend to their conclusions cannot be deductive either.

On the other hand, there seems to be a problem in finding some non-deductive kind of justification. For instance, it might be natural to argue that it is justifiable to employ induction because experience tells us that inductive arguments are, by and large, successful. We inevitably employ inductive arguments every day. Indeed all our actions depend for their success on the world continuing to behave as it has done. Every morsel we eat and every step we take reflect our reliance on food continuing to nourish and the ground continuing to remain firm. So it would seem that our very existence is testimony to a high degree of reliability among inductive arguments. But, to argue that past success in inductive reasoning and inductive behaviour is a reason for being confident that future inductive reasoning and behaviour will be successful, is itself to argue in an inductive manner. It is to use induction to justify induction. In that case the argument is circular.

Hume's point can be put another way. How could we improve our inductive argument so that its justification would not be in question? One way to try would be to add an additional premise that turns the argument into a deductive one. For instance, along with the premise "all observed cases of colitis have been accompanied by anaemia" we may add the premise "unobserved cases resemble observed cases" (let us call this *the uniformity premise*). From the two premises we may deduce that unobserved colitis sufferers are anaemic too. We are now justified in believing our conclusion, but only if we are justified in believing the uniformity premise, that unobserved cases resemble observed ones. The uniformity premise is itself clearly a generalization. There is evidence for it – cases where something has lain unobserved and has at length been found to resemble things already observed. The lump of ice I took out of the freezer an hour ago had never hitherto been observed to melt at room temperature (16°C in this case), but it did melt, just as all previous ice cubes left at such a temperature have been observed to do. Experience tells in favour of the uniformity premise. But such experience can only justify the uniformity premise if we can expect cases we have not experienced to resemble the ones we have experienced. That expectation is just the same as the uniformity premise. So we cannot argue for the uniformity premise without assuming the uniformity premise. But without the uniformity premise – or something like it – we cannot justify the conclusions of our inductive arguments. Induction cannot be made to work without also being made to be circular. (Some readers may wonder whether the uniformity premise is in any case true. While our expectations are often fulfilled they are frequently thwarted too. This is a point to which we shall return in Chapter 5.)

If the premises of an inductive argument do not entail their conclusions, they can give them a high degree of probability. Might not Hume's problem be solved by probabilistic reasoning? I shall examine this claim in detail in Chapter 6. But we can already see why this cannot be a solution. Consider the argument "all observed colitis suffers are anaemic, therefore it is likely that all colitis sufferers are anaemic". This argument is not deductively valid, so we cannot justify it in the way we justify deductive arguments. We could justify it by saying that arguments of this kind have been successful in many cases and are likely to be successful in this case too. But to use this argument is to use an argument of the form we are trying to justify. We could also run the argument with a modified uniformity premise: a high proportion of unobserved cases resemble observed cases. But again, we could only justify the modified uniformity premise with an inductive argument. Either way the justification is circular.

Goodman's problem

There is another problem associated with induction, and this one is much more recent in origin. It was discovered by the American philosopher Nelson Goodman. Hume's problem concerns the question: Can an inductive argument give us knowledge (or justification)? If the argument employed is right, it would seem that those arguments we would normally rely upon and expect to give us knowledge of the future or of things we have not experienced, cannot really give us such knowledge.

This is a disturbing conclusion, to find that our most plausible inductive arguments do not give us knowledge. Even if there is a solution, there is the further problem of identifying those arguments that are even plausible as inductive arguments. A deductive argument is one the premises of which entail its conclusion. I have said little about inductive arguments in general, except that they do not entail their conclusions. However, I did ascribe some structure to what I called Humean inductions. One example was "all observed patients with colitis have anaemia, therefore all colitis sufferers have anaemia". Another example beloved of philosophers is "all observed emeralds are green, therefore all emeralds are green". In these cases we are generalizing from what we have observed of a certain kind of thing, to all of that kind, both observed and unobserved, past, present, and future. We can express this in a pattern or general format for Humean induction:

all observed Fs are Gs;
therefore
all Fs are Gs.

Goodman's problem, the "new riddle of induction" as it is also known, suggests that this format is too liberal. It allows in too many arguments that we would regard as wildly implausible. Worse, it seems to show that for every inductive argument we regard as plausible there are many implausible arguments based on precisely the same evidence. Let us see how this works.

Goodman introduces a new term *grue*. *Grue* is defined thus:

X is grue = either X is green and observed before midnight on 31 December 2000
or X is blue and not observed before midnight on 31 December 2000

Such an unintuitive and novel concept is perhaps not easy to grasp, so let us look at a few cases. A green gem that has already been dug up and examined is grue. A green gem that remains hidden under ground until 2001 is not grue, but a hidden blue gem first examined in 2001 is grue. Were the blue gem dug up in 1999 it would not be grue.

All observed emeralds are both green and grue. To see that they are all grue, consider that since they are all observed (and I am writing in 1997) they are observed before midnight on 31 December 2000. They are also all green. So they satisfy the definition of grue (by satisfying the first of the two alternative clauses).

Returning to the general format for an inductive argument, we should have an inductive argument of the form:

all observed emeralds are grue;
therefore
all emeralds are grue.

The conclusion is that all emeralds are grue. In particular, emeralds that are first dug up and examined in 2001 are grue. From the definition of grue, this means they will be blue. So the format of our inductive argument allows us to conclude that emeralds first examined in 2001 are blue. And it also allows us to conclude, as we already have, that these emeralds will be green.

Predicates such as "grue" are called *bent* predicates, as their meanings seem to involve a twist or change; they are also called *non-projectible*, because our intuitions are that we cannot use them for successful prediction. Bent predicates enable us to create, for every credible inductive argument, as many outrageous inductions as we like. So not only can we argue that all emeralds are green and argue that they are grue, we can also argue that they are gred and that they are grellow. A standard induction would suggest that water at atmospheric pressure boils at 100°C. But we can form a Goodmanesque induction the conclusion of which is that water breezes at 100°C (where the definition of "breezes" implies that water observed for the first time in the next millennium will freeze at 100°C). In addition there are arguments that children born in the next century will be immortal and that the Earth's rotation will reverse so that the Sun will rise in the west.

The problem with this is not merely that these conclusions seem thoroughly implausible. Rather, it looks as if any claim about the future can be made to be a conclusion of an inductive argument from any premises about the past, as long as we use a strange enough grue-like predicate. And, if anything can be made out to be the conclusion of an inductive argument, then the very notion of inductive argument becomes trivial. So how do we characterize the reasoning processes of science if their central notion is trivial?

Philosophers have proposed a wide range of answers and responses to Goodman's problem, and I will myself make a suggestion in Chapter 3. But one response that will not do is to dismiss the problem as contrived. First, this begs the question: Why is it contrived? What distinguishes pseudo-inductive arguments with contrived concepts like "grue" from genuine inductions using "green"? Secondly, we can think of cases that are less contrived. Take someone who knows about trees only that some are evergreens and some are deciduous. If they were to observe beech trees every day over one summer they would have evidence that seems to support both the hypothesis that beech trees are deciduous and the hypothesis that they are evergreens. Here neither option is contrived. But we know this only because we have experienced both. (The idea of deciduousness might appear contrived to someone who had lived all their lives in pine forests.)

Goodman's problem appears to be at least as fundamental as Hume's, in the following sense. Even if we had a solution to Hume's problem, Goodman's problem would appear to remain. Hume argues that no conclusion to an inductive argument can ever amount to knowledge. Say we thought that there were an error in Hume's reasoning, and that knowledge from inductive arguments is not ruled out. It seems we would still be at a loss to say which arguments these are, as nothing said so far could distinguish between a genuine inductive one and a spurious one involving bent predicates.

Representation and reason

When discussing Judge Overton's opinion regarding creationism, I stressed that he was not trying to judge whether creationism, or evolution, or something else is true, let alone whether anyone could rightly claim to know which of these is true. Rather, he was, in part, trying to judge what it is that makes a claim to be scientific. Roughly speaking, there were two sorts of relevant answer. The first dealt with the subject matter of science: laws of nature, natural explanation, natural kinds (if any). The second concerned the sorts of attitude and approach a scientist has towards a scientific theory: proportioning belief to the strength of evidence, avoiding dogma, being open to the falsifiability of a theory. These two kinds of answer correspond to the structure of this book, the remaining chapters of which are divided into two sections: *representation* and *reason*.

The first part is called *representation* because (so I claim) the basic aim of science is to provide an accurate representation of the world – what sorts of things are in it, how they interact, what explains what, and so on. In the previous paragraph I mentioned certain very general concepts that are used in the scientific representation of the world, and I will ask what is meant by these. We start in Chapter 1 with the concept *law of nature*, and then move on in Chapter 2 to the notion of *explanation*, which will allow us also to look, albeit briefly, at *causes*. Chapter 3 deals with the contentious question of whether the concept of a *natural kind* is needed in science and, if so, how it should be understood. In Chapter 4 I will pause to consider whether I was right to claim that science aims at representation – as opposed merely to being a tool for prediction. One reason for taking the latter view is the claim that we cannot have knowledge of the unobservable parts of the world. In which case it would be unreasonable for science to aim at representing the world because it cannot achieve this.

This leads to the subject matter of the second part of the book, *reason*. Here the question is how much can scientific reasoning tell us about the world. In Chapter 5 I start with the sceptical position that it can tell us nothing – scientific knowledge of a general kind is impossible. We will look at various proposals for solutions to Hume's problem. In particular, I will assess Popper's attempt to avoid the problem by giving an account of science that does not depend on induction. In Chapter 6 I look at the contribution made by probability to scientific reasoning, and whether some mathematical theory can provide an *a priori* basis for proportioning belief to the evidence. I do think that an answer to Hume's problem can be found, and I present this in Chapter 7. There we will look at developments in modern epistemology that have provided a fruitful way of addressing not only the problem of induction but also other problems in the philosophy of science, such as the so-called "theory-ladenness of observation". In the final chapter, I address the question of whether there is such a thing as the scientific method. I come to the conclusion that there is not, and consequently ask whether, without "the scientific method" we can explain the progress of science, and indeed whether there is any.

Further reading

An extract from Judge Overton's judgment is printed in Joel Feinberg (ed.), *Reason and responsibility*. The case for creationism is made by Henry Morris in *Scientific creationism*; and a philosopher's case against it is Philip Kitcher's *Abusing science*. An excellent discussion of what scientists do and why, which I recommend, even to scientists, is Leslie Stevenson and Henry Byerly's *The many faces of science*. Another readable book which discusses several case studies in the history of science is *The golem* by Harry Collins and Trevor Pinch. David Hume spells out the problem for induction in his *Enquiry concerning human understanding* (especially Section IV and Section VII, Part ii), and Nelson Goodman introduces his puzzle about grue in *Fact, fiction and forecast*.

Part I

Representation

Chapter 1

Laws of nature

The notion of a law of nature is fundamental to science. In one sense this is obvious, in that much of science is concerned with the discovery of laws (which are often named after their discoverers – hence Boyle's law, Newton's laws, Ostwald's law, Mendel's laws, and so on). If this were the only way in which laws are important, then their claim to be fundamental would be weak. After all, science concerns itself with much more than just the uncovering of laws. Explaining, categorizing, detecting causes, measuring, and predicting are other aims of scientific activity, each quite distinct from law-detecting. A claim of this book is that laws are important because each of these activities depends on the existence of laws. For example, take Henry Cavendish's attempt to measure the gravitational constant G using the torsion balance. If you were to have asked him what this constant is, he would have told you that it is the ratio of the gravitational force between two objects and the product of their masses divided by the square of their separation. If you were then to ask why this ratio is constant, then the answer would be that it is a law of nature that it is so. If there were no law of gravitation, then there would be no gravitational constant; and if there were no gravitational constant there would be nothing that counts as the measurement of that constant. So the existence and nature of a law was essential to Cavendish's measuring activities. Somewhat contentiously, I think that the notion of explanation similarly depends upon that of law, and even more contentiously I think that causes are also dependent on laws. A defence of these views must wait until the next chapter. I hope it is clear already that laws are important, a fact which can be seen by considering that if there were no laws at all the world would be an intrinsically chaotic and random place in which science would be impossible. (That is assuming that a world without laws could exist – which I am inclined to think doubtful, since to be any sort of thing is to be subject to some laws.) Therefore, in this chapter we will consider the question: *What is a law of nature?* Upon our answer, much that follows will depend.

Before going on to discuss what laws of nature are, we need to be clear about what laws are not. Laws need to be distinguished from *statements* of laws and from our *theories* about what laws there are. Laws are things in the world which we try to discover. In this sense they are facts or are like them. A statement of a law is a linguistic item and so need not exist, even though the corresponding law exists (for instance if there were no people to utter or write statements). Similarly, theories are human creations but laws of nature are not. The laws are there whatever we do – one of the tasks of the scientist is to speculate about them and investigate them. A scientist may come up with a theory that will be false or true depending on what the laws actually are. And, of course, there may be laws about which we know nothing. So, for instance, the law of universal gravitation was discovered by Newton. A statement of the law, contained in his theory of planetary motion, first appeared in his *Principia mathematica*. In the following we are interested in what laws are – that is, what sort of thing it was that Newton discovered, and not in his statement of the law or his theories containing that statement.[8] Furthermore, we are not here asking the question: Can we know the existence of any laws? Although the question "What is an X?" is connected to the question "How do we know about Xs?", they are still distinct questions. It seems sensible to start with the former – after all, if we have no idea what Xs are, we are unlikely to have a good answer to the question of how we know about them.

Minimalism about laws – the simple regularity theory

In the Introduction we saw that the function of (Humean) induction is to take us from observations of particular cases to generalizations. It is reasonable to think that this is how we come to know laws – if we can come to know them at all. Corresponding to a true inductive conclusion, a true generalization, will be a *regularity*. A regularity is just a general fact. So one inductive conclusion we might draw is that all emeralds are green, and another is that all colitis patients suffer from anaemia. If these generalizations are true, then it is a fact that each and every emerald is green and that each and every colitis sufferer is anaemic. These facts are what I have called regularities. The first view of laws I shall consider is perhaps the most natural one. It is that laws are just regularities.

This view expresses something that I shall call *minimalism* about laws. Minimalism takes a particular view of the relation between a law and its instances. If it is a law that bodies in free fall near the surface of the Earth accelerate towards its centre at 9.8 ms^{-2}, then a particular apple falling to the ground and accelerating at 9.8 ms^{-2} is an *instance* of this law. The minimalist says that the law is essentially no more than the collection of all such instances. There have been, and will be, many particular instances, some observed but most not, of objects accelerating towards the centre of the Earth at this rate. The law, according to the minimalist, is simply the regular occurrence of its instances. Correspondingly, the statement of a law will be the generalization or summary of these instances.

Minimalism is an expression of *empiricism*, which, in broad terms, demands that our concepts be explicable in terms that relate to our experiences. Empiricist minimalism traces its ancestry at least as far back as David Hume. By defining laws in terms of regularities we are satisfying this requirement (as long as the facts making up the regularities are of the sort that can be experienced).

Later we shall come across an approach to laws that is not empiricist.

The simplest version of minimalism says that laws and regularities are the same. This is called the *simple regularity theory* (SRT) of laws.

SRT: It is a law that Fs are Gs *if and only if* all Fs are Gs.

While the SRT has the merit of simplicity it suffers from the rather greater demerit that it is false. If it were true, then all and only regularities would be laws. But this is not the case.

The most obvious problem is that the existence of a simple regularity is not sufficient for there to be a corresponding law, i.e. there are simple regularities that are not laws. A criticism is also made that it is not even necessary for there to be a regularity for the corresponding law to exist. That is, there are laws without the appropriate regularities.

Regularities that are not laws

I will start with the objection that being a regularity is not sufficient for being a law. Consider the following regularities:

(a) All persisting lumps of pure gold-195 have a mass less than 1,000 kg.
(b) All persisting lumps of pure uranium-235 have a mass of less than 1,000 kg.[9]

Both (a) and (b) state true generalizations. But (a) is accidental and (b) is law-like. It is no law that there are no lumps of the pure isotope of gold – we could make one if we thought it worth our while. However, it is a law that there are no such lumps of uranium-235, because 1,000 kg exceeds the critical mass of that isotope (something less than a kilogram) and so any such lump would cause its own chain reaction and self-destruct. What this shows is that there can be two very similar looking regularities, one of which is a law and the other not.

This is not an isolated example. There are an indefinite number of regularities that are not laws. Take the generalization: all planets with intelligent life forms have a single moon. For the sake of argument, let us imagine that the Earth is the only planet in the universe with intelligent life and that there could exist intelligent life on a planet with no moons or more than one. (For all I know, these propositions are quite likely to be true.) Under these circumstances, the above generalization would be true, even though there is only one instance of it. But it would not be a law; it is just a coincidence. The point here is that the SRT does not distinguish between genuine laws and mere coincidences. What we have done is to find a property to take the place of F which has just one instance and then we take any other property of that instance for G. Then "All Fs are Gs" will be true. And, with a little thought, we can find any number of such spurious coincidental regularities. For most things that there are we could list enough of their general properties to distinguish one thing from everything else. So, for instance, with a person, call her Alice, we just list her hair colour, eye colour, height, weight, age, sex, other distinguishing features, and so on in enough detail that only Alice has precisely those qualities. These qualities we bundle together as a single property F. So only Alice is F. Then choose some other property of Alice (not necessarily unique to her), say the fact that she plays the oboe. Then we will have a true generalization that all people who are F (i.e. have fair hair, green eyes, are 1.63 m tall, weigh 59.8 kg, have a retroussé nose, etc.) play the oboe. But we do not want to regard this as a law, since the detail listed under F may have nothing whatsoever to do with an interest in and talent for playing the oboe.

The minimalist who wants to defend the SRT might say that these examples look rather contrived. First, is it right to bundle a lot of properties together as one property F? Secondly, can just one instance be regarded even as a regularity? (If it is not a regularity then it will not be a counter-instance.) However, I do not think that the minimalist can make much headway with these defences.

To the general remark that the examples look rather contrived, the critic of the SRT has two responses. First, not all the cases are contrived, as we can see from the gold and uranium example. We can find others. One famous case is that of Bode's "law" of planetary orbits. In 1772, J. E. Bode showed that the radii of known planetary orbits fit the following formula: $0.4 + 0.3 \times 2^n$ (measured in astronomical units) where $n = 0$ for Venus, 1 for the Earth, 2 for Mars, and so on, including the minor planets. (Mercury could be included by ignoring the second term, its orbital radius being 0.4 astronomical units.) Remarkably, Bode's law was confirmed by the later discovery of Uranus in 1781. Some commentators argued that the hypothesis was so well confirmed that it achieved the status of a law, and consequently ruled out speculation concerning the existence of a possible asteroid between the Earth and Mars. Bode's "law" was eventually refuted by the observation of such an asteroid, and later by the discovery of the planet Neptune, which did not fit the pattern. What Bode's non-law shows is that there can be remarkable uniformities in nature that are purely coincidental. In this case the accidental nature was shown by the existence of planets not conforming to the proposed law. But Neptune, and indeed Pluto too, might well have had orbits fitting Bode's formula. Such a coincidence would still have been just that, a coincidence, and not sufficient to raise its status to that of a law.

Secondly, the critic may respond that the fact that we can contrive regularities is just the point. The SRT is so simple that it allows in all sorts of made-up regularities that are patently not laws. At the very least the SRT will have to be amended and sharpened up to exclude them. For instance, taking the first specific point, as I have stated it the SRT does not specify what may or may not be substituted for F and G. Certainly it is an important question whether compounds of properties are themselves also properties. In the Alice example I stuck a whole lot of properties together and called them F. But perhaps sticking properties together in this way does always yield a new property. In which case we might want to say that only uncompounded properties may be substituted for F and G in the schema for the SRT.

However, amending the SRT to exclude Fs that are compound will not help matters anyway, for two reasons. First, there is no reason why there should not be uncompounded properties with unique instances. Secondly, some laws do involve compounds – the gas laws relate the pressure of a gas to the compound of its volume and temperature. To exclude regularities with compounds of properties would be to exclude a regularity for which there is a corresponding law. To the second point, that the regularities constructed have only one instance, one rejoinder must be this: why cannot a law have just one instance? It is conceivable that there are laws the only instance of which is the Big Bang. Indeed, a law might have no instances at all. Most of the transuranium elements do not exist in nature and must be produced artificially in laboratories or nuclear explosions. Given the difficulty and expense of producing these isotopes and because of their short half-lives it is not surprising that many tests and experiments that might have been carried out have not been. Their electrical conductivity has not been examined, nor has their chemical behaviour. There must be laws governing the chemical and physical behaviour of these elements under circumstances which have never and never will arise for them. There must be facts about whether nobelium-254, which is produced only in the laboratory, burns in oxygen and, if so, what the colour of its flame is, what its oxide is like, and so forth; these facts will be determined by laws of nature, laws which in this case have no instances.

So we do not want to exclude something from being a law just because it has few instances, even just one instance, or none at all. At the same time the possibility of instanceless laws raises another problem for the SRT similar to the first. According to the SRT empty laws will be empty regularities – cases of "all Fs are Gs" where there are no Fs. There is no problem with this; it is standard practice in logic to regard all empty generalizations as trivially true.[10] What is a problem is how to distinguish those empty regularities that are laws from all the other empty regularities. After all, a trivial empty regularity exhibits precisely as much regularity as an empty law.

Let us look at a different problem for the SRT. This concerns functional laws. The gas law is an example of a functional law. It says that one magnitude – the pressure of a body of gas – is a function of other magnitudes, often expressed in a formula such as:

$$P = k\frac{T}{V}$$

where P is pressure, T is temperature, V is volume, and k is a constant. In regarding this as a law, we believe that T and V can take any of a continuous range of values, and that P will correspondingly take the value given by the formula. Actual gases will never take all of the infinite range of values allowed for by this formula. In this respect the formula goes beyond the regularity of what actually occurs in the history of the universe. But the SRT is committed to saying that a law is just a summary of its instances and does not seek to go beyond them. If the simple regularity theorist sticks to this commitment, the function ought to be a partial or gappy one, leaving out values that are not actually instantiated. Would such a gappy "law" really be a law? One's intuition is that a law should cover the uninstantiated values too. If the SRT is to be modified to allow filling in of the gaps, then this needs to be justified. Furthermore, the filling in of the gaps in one way rather than another needs justification. Of course there may well be a perfectly natural and obvious way of doing this, such as fitting a simple curve to the given points. The critic of the SRT will argue that this is justified because this total (non-gappy) function is the best explanation of the instantiated values. But the simple regularity theorist cannot argue in this way, because this argument accepts that a law is something else other than the totality of its instances. As far as what actually occurs is concerned, one function which fits the points is as good as any other. From the SRT point of view, the facts cannot decide between two such possible functions. But, if the facts do not decide, we cannot make an arbitrary choice, say choosing the simplest for the sake of convenience, as this would introduce an arbitrary element into the notion of lawhood. Nor can we allow all the functions that fit the facts to be laws. The reason for this is the same as the reason why we cannot choose one arbitrarily and also the same as the reason why we cannot have all empty generalizations as laws. This reason is that we expect laws to give us determinate answers to questions of what would have happened in *counterfactual* situations – that is situations that did not occur, but might have.

Laws and counterfactuals

Freddie's car is black and when he left it in the sun it got hot very quickly. The statement "Had Freddie's car been white, it would have got hot less quickly" is an example of a counterfactual statement. It is not about what did happen, but what would have happened in a possible but not actual (a counter-to-fact) situation (i.e. Freddie's car being white rather than black). The counterfactual in question is true. And it is true because it is a law that white things absorb heat less rapidly than black things. Laws support counterfactuals.

We saw above that every empty regularity will be true and hence will be a law, according to the SRT. This is an undesirable conclusion. Counterfactuals help us see why. Take some property with no instances, F. If we allowed all empty regularities to be laws we would have both law 1 "it is a law that Fs are Gs" and law 2 "it is a law that Fs are not-Gs". What would have happened if *a*, which is not F, had been F? According to law 1, *a* would have been G, while law 2 says *a* would have been not-G. So they cannot both really be laws. Similarly, we cannot have both of two distinct functions governing the same magnitudes being laws, even if they agree in their values for actual instances and diverge only for non-actual values. For the two functional laws will contradict one another in the conclusion of the counterfactuals they support when we ask what values would *P* have taken had *T* and *V* taken such-and-such (non-actual) values.

Counterfactuals also underline the difference between accidental and nomic regularities. Recall the regularities concerning very large lumps of gold and uranium isotopes. There are no such lumps of either. In the case of uranium-235, there could not be such lumps, there being a law that there are no such lumps. On the other hand, there is no law concerning large lumps of gold, and so there could have been persisting 2,000 kg lumps of gold-195. In this way counterfactuals distinguish between laws and accidents.

Some philosophers think that the very fact that laws support counterfactuals is enough to show the minimalist to be wrong (and the SRT supporter in particular). The reasoning is that counterfactuals go beyond the actual instances of a law, as they tell us what would have happened in possible but non-actual circumstances. And so the minimalist must be mistaken in regarding laws merely as some sort of summary of their actual instances. This argument seems powerful, but I think it is not a good line for the anti-minimalist to pursue. The problem is that counterfactuals are not any better understood than laws, and one can argue that our understanding of counterfactuals is dependent on our notion of law or something like it, in which case corresponding to the minimalist account of laws will be a minimalist account of counterfactuals.[11] You can see that this response is plausible by considering that counterfactuals are read as if there is a hidden clause, for instance "Freddie's car would have got hot less quickly had it been white *and everything else and been the same as far as possible*". (Which is why one cannot reject the counterfactual by saying that had Freddie's car been white, the Sun might not have been shining.) The clause which says that everything should be the same as far as possible requires among other things, like the weather being the same, that the laws of nature be the same. So one can say that laws support counterfactuals only because counterfactuals implicitly refer to laws. Counterfactuals therefore have nothing to tell us about the analysis of laws. Consider the fact that laws do not support all counterfactuals – particularly those counterfactuals relating to the way things would be with different laws. For instance one could ask how quickly would two things have accelerated towards one another had the gravitational constant G been twice what it is. The actual law of gravitation clearly does not support the correct answer to this counterfactual question.

Laws that are not regularities – probabilistic laws

So far there is mounting evidence that being a simple regularity is not sufficient for being a law. There are many regularities that do not constitute laws of nature:

(a) accidental regularities
(b) contrived regularities
(c) uninstantiated trivial regularities
(d) competing functional regularities.

The natural response on the part of the minimalist is to try to amend the SRT by adding further conditions that will reduce the range of regularities which qualify as laws. So the original idea is maintained, that a law is a regularity, and with an appropriate amendment it will now be that laws are a certain sort of regularity, not any old regularity.

This is the line I will examine shortly. But before doing so I want to consider an argument which suggests that being a regularity is not even sufficient for being a law. That is, there are laws that are not simple regularities. If this line of thinking were correct, then it would be no good trying to improve the SRT by adding extra conditions to cut down the regularities to the ones we want, as this would still leave out some laws that do not have a corresponding regularity.

The problem concerns probabilistic laws. Probabilistic laws are common in nuclear physics. Atomic nuclei as well as individual particles are prone to decay. This tendency to decay can be quantified as the probability that a nucleus will decay within a certain period. (When the probability is one-half, the corresponding period of time is called the *half-life*.) So a law of nuclear physics may say that nuclei of a certain kind have a probability p of decaying within time t. What is the regularity here? The SRT, as it stands, has no answer to this question. But an answer, which satisfies the minimalist's aim of portraying laws as summaries of the individual facts, is this. The law just mentioned will be equivalent to the fact that, of all the relevant particles taken together, a proportion p will have decayed within t. (Note that we would find out what the value of p is by looking at the proportion that decays in observed samples. Another feature that might be included in the law is *resiliency*, i.e. that p is the proportion which decays in all appropriate subpopulations, and not just the population as a whole.)

The problem arises when we consider each of the many particles individually. Each particle has a probability p of decaying in time t. This is perfectly consistent with the particle decaying well before t or well after t. So the law allows for any particle to decay only after time t^*, which is later than t. And what goes for one particle goes for all. Thus the law is consistent with all the particles decaying only after t^*, in which case the proportion decaying by t is zero. This means that we have a law the instances of which do not form the sort of regularity the minimalist requires. The minimalist requires the proportion p to decay by t. We would certainly expect that. But this is by no means necessary. While it is extremely unlikely that no particle will decay until after *t*, it is not impossible. Another, less extreme, case would be this. Instead of the proportion *p* decaying within *t*, a very slightly smaller proportion than *p* might decay. The chance of this is not only greater than zero, i.e. a possibility, but may even be quite high.

The minimalist's guiding intuition is that the existence and form of a law is determined by its instances. Here we have a case where our intuitions about laws allow for a radical divergence between the law and its instances. And so such a law is one which could not be a regularity of the sort required by the minimalist. This would appear to be a damning argument against minimalists, not even allowing for an amendment of their position by the addition of extra conditions. Nonetheless, I think the argument employed against the minimalist on this point could be seen as simply begging the question. The argument starts by considering an individual particle subject to a probabilistic law. The law might be that such particles have a half-life of time *t*. The argument points out that this law is consistent with the particle decaying at time t^* after *t*. This is all true. The argument then proceeded to claim that any collection could, therefore, be such that all its particles decay after *t*, including the collection of all the particles over all time.

The form of this argument is this: what is possible for any individual particle is possible for a collection of particles; or, more generally, if it is possible that X, and it is possible that Y, and it is possible that Z, and so on, then it is possible that X and Y and Z, etc. This form of argument certainly is not logically valid. Without looking outside, I think to myself it is possible that it is raining and it is possible that it is not raining, but I certainly should not conclude that it is possible that it is both raining and not raining. More pertinent to this issue is another counter-example. It is possible for one person to be taller than average, but it is not possible for everyone to be taller than average. This is because the notion of an average is logically related to the properties of a whole group. What this suggests is that the relevant step in the argument against the minimalist could only be valid if there is a logical gap between a law and the collection of its instances. But this is precisely what the minimalist denies. For, on the contrary, the minimalist claims that there is no gap; rather there is an intimate logical link in that the probabilistic law is some sort of sophisticated averaging out over its instances. So it turns out that the contentious step in the argument is invalid or, at best, question begging against the minimalist.

Still, even if the argument is invalid, it does help alert our intuitions to an implausibility with minimalism. We may not be able to prove that there can be a gap between the chance of decay that each atom has and the proportion of particles actually decaying – but the fact that such a gap does *seem* possible is evidence against the minimalist.

The systematic account of laws of nature

We have seen so far that not all regularities are also laws, though it is possible for the minimalist to resist the argument that not all laws are regularities. What the minimalist needs to do is to find some further conditions, in addition to being a regularity, which will pare down the class of regularities to just those that are laws.

Recall that the minimalist wants laws simply to generalize over their instances. Law statements will be summaries of the facts about those instances. If this is right we should not expect every regularity to be a law. For we can summarize a set of facts without having to mention every regularity that they display, and an efficient summary would only record sufficient regularities to capture all the facts in which we are interested. Such a summary would be *systematic*; rather than being a list of separate regularities, it would consist of a shorter list of interlocking regularities, which together are enough to encapsulate all the facts. It would be as simple as possible, but also as strong as possible, capturing as many facts and possible facts as it can.

What I mean here by "capturing" a fact is the subsuming of that fact under some law. We want it to be that laws capture all and only their instances. So if the fact is that some object *a* has the property G, then this would be captured by a law of the form all Fs are Gs (where *a* is F). For example, if the fact we are interested in is the explosive behaviour of a piece of lithium, then this fact can be captured by the law that lithium reacts explosively on contact with water (if in this case it was in contact with water). If the fact is that some magnitude *M* has the value *m* then the fact is captured by a law of the form $M = f(X, Y, Z, \ldots)$ (where the value of X is x, the value of Y is y, etc., and $m = f(x, y, z, \ldots)$. An example of this might be the fact that the current through a circuit is 0.2 A. This fact can be captured by Ohm's law $V = IR$, if for instance there is a potential difference of 10 V applied to a circuit that has a resistance of 50 Ω. As we shall see in the next chapter, this is tantamount to saying that the facts in question have explanations. If a fact has an explanation according to one law, we do not need another, independent law to explain the same fact. Let it be the case that certain objects, *a*, *b*, *c*, and *d* are all G. Suppose that they are also all F, and also the only Fs that there are. We might think then that it is a law that Fs are Gs and that this law explains why each of the objects is G. But we could do without this law and the explanations it furnishes if it is better established that there is a law that Hs are Gs and *a* is H, and that there is a law that Js are Gs and that *b* is J, and that there is a law that Ks are Gs and that *c* is K, and so on. In this case the proposed law that Fs are Gs is redundant. We can instead regard the fact that all Fs are Gs as an accidental regularity and not a law-like one. If we symbolize "*a* is F" by "F*a*", then as Figure 1.1 shows, we could organize the facts into an economical system with three rather than four laws.

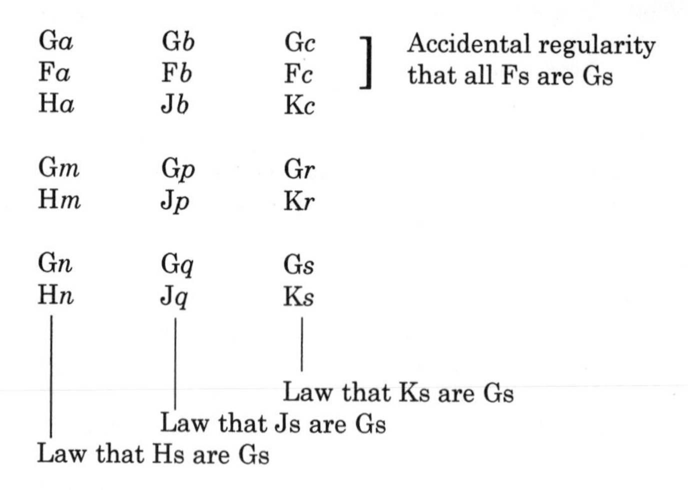

Figure 1.1

This organizing of facts into an economical system allows us to distinguish between accidental regularities and laws. This was one of the problems facing the SRT. The systematic approach does better. It also easily accommodates the other problem case, functional laws. The simplest way of systematizing a collection of distinct points on a graph is to draw the simplest line through them. And we will only want one line, as adding a second line through the same points will make the system much less simple without adding to the power of the system, because the first line already captures the facts we are interested in. Thus in a systematic version of minimalism we will have non-gappy functional laws that are unique in their domain.

These are the virtues of the account of laws first given by Frank Ramsey and later developed by David Lewis.[12] The idea is that we regard the system as axiomatized, that is to say we boil the system down to the most general principles from which the regularities that are laws follow. A collection of facts may be axiomatized in many (an infinite number of) ways. So the appropriate axiomatization will be that which, as discussed above, is as simple as possible and as strong as possible. The strength of a proposition can be regarded as its informativeness. So, considering (a) all emeralds are green, (b) all emeralds are coloured, and (c) all South American emeralds are green, (a) is more informative and so stronger than both (b) and (c).

Formally, the systematic account says:

> A regularity is a law of nature if and only if it appears as a theorem or axiom in that true deductive system which achieves a best combination of simplicity and strength.

Remember that this is a "God's eye" view, not ours, in the sense that we do not know all the facts that there are and so do not know in that direct way what the best axiomatization of them is. What we are saying is that this is what a law of nature is. Our ignorance of the best axiomatization of everything is identical to our ignorance of the laws of nature. Of course we are not entirely ignorant of them. If science is any good we know some of them, or approximations to them at any rate. And it can be said in support of Ramsey and Lewis, that the sorts of thing we look for in a theory which proposes a law of nature are precisely what they say something ought to have in order to be a law. First we look to see whether it is supported by evidence in the form of instances, i.e. whether it truly is a regularity. But we will also ask whether it is simple, powerful, and integrates with the other laws we believe to exist.

The systematic view is not entirely without its problems. First, the notion of *simplicity* is important to the systematic characterization of law, yet simplicity is a notoriously difficult notion to pin down. In particular, one might think that simplicity is a concept which has a significant subjective component to it. What may appear simple to one person might look rather complex to another. In which case the concept of law would also have a subjective element that conflicts with our intuition that laws are objective and independent of our perspective. Another way of looking at the problem of simplicity is to return to Goodman's puzzle about the concept "grue". The fact that we can replace the simple looking "X is grue" by the complex looking "either X is green and observed before midnight on 31 December 2000 or X is blue and not observed before midnight on 31 December 2000", and vice versa, shows that we cannot be sure of depending merely on linguistic appearance to tell us the difference. The simple law that emeralds are green appears to be the same as the complex law that emeralds are grue, if observed before midnight on 31 December 2000, or bleen if not observed before midnight on 31 December 2000. Without something like a solution to Goodman's problem we have no reason to prefer one set of concepts to another when trying to pin down the idea of simplicity.

The second problem with the systematic view is that as I have presented it, i.e. it presumes there is precisely one system that optimally combines strength and simplicity. For a start it is not laid down how we are supposed to weigh simplicity and strength against one another. We could add more laws to capture more potential facts and thus gain strength, but with a loss in simplicity. Alternatively, we might favour a simpler system that has less strength. Again there is the suspicion that the minimalist's answer to this problem may be objectionably subjective. Even if there is a clear and objective way of balancing these two, it may yet turn out that two or more distinct systems do equally well. So which are our laws? Perhaps something is a law if it appears in any one of the optimal systems. But this will not do, because the different systems might have conflicting laws that lead to incompatible counterfactuals. On the other hand, we may be more restrictive and accept as laws only those regularities that appear in all the optimal systems.[13] However, we may find that there are very few such common regularities, and perhaps none at all.

Basic laws and derived laws

It is a law-like and true (almost) generalization that all objects of 10 kg when subjected to a force of 40 N accelerate at 4 m s^{-2}. But this generalization would not feature as an axiom of our system, because it would be more efficient to have the generalization that the acceleration of a body is equal to the ratio of the resultant force acting upon it and its mass. By having that generalization we can dispense with the many generalizations that mention specific masses, forces, and accelerations. From the overarching generalization these more specific ones may be derived. That is to say, our system will be axiomatic; it will comprise the smallest set of laws from which it is possible to deduce all other laws. This shows an increase in simplicity – few generalizations instead of many – and in strength, as the overarching generalization will have consequences not entailed by the collection of specific generalizations. In the example there may be some values of the masses, forces, and accelerations for which there are no specific generalizations, just because nothing ever both had that mass and was subject to that force.

This illustrates a distinction between more fundamental laws and those derived from them. Most basic of all are those laws that form the axioms of the optimal system. If the laws of all the sciences can be reduced to laws of physics, as some argue is true, at least for chemistry, then the basic laws will be the fundamental laws of physics. If, on the other hand, there is some field the laws of which are not reducible, then the basic laws will include laws of that field. It might be that we do not yet actually know any basic laws – all the laws we are acquainted with would then be derived laws.

Laws and accidents

I hope that it should now appear that the systematic view of laws is a considerable improvement on the SRT. While it still has problems, the account looks very much as if it is materially adequate. An analysis of a concept is *materially adequate* if it is true that were anything to be a case of the concept (the *analysandum* – the thing to be analysed) it would also be a case of the analysis, and vice versa. So "female fox" is a materially adequate analysis of "vixen" only when it is true that if something were a female fox it would also be a vixen, and if something were a vixen it would also be a female fox. The systematic account looks to be materially adequate because the regularities that are part of or consequences of the optimal system are precisely the regularities we would identify as nomic regularities (regularities of natural law).

In this section I want to argue that this is an illusion, that it is perfectly possible that systematic regularities fail to correspond to the laws there are. And in the next section we will see that even if, as a matter of fact, laws and systematic regularities coincide, this is not because they are the same things. The point will be that there are certain things we use laws for, in particular for *explaining*, for which we cannot use regularities, however systematic. And towards the end of this chapter I shall explain what sort of connection there is between laws and systematic regularities, and why therefore the systematic account *looks* as if it is materially adequate.

Earlier on we looked at the argument that probabilistic laws were a counter-example to minimalism. The idea was that there could be a divergence between the law and the corresponding regularity. Although the argument is not valid, I suggested that our intuitions should be against the minimalist. More generally, I believe that our intuitions suggest that there could be a divergence between the laws there are and the (systematic) regularities there are.

Returning to the simple regularity theory for a moment, one reason why this theory could not be right is that there could be a simple regularity that is merely accidental. The fact that there is regularity is a coincidence and nothing more. These instances are not tied together by a law that explains them all. To some extent this can be catered for by the systematic account. For, if the regularity is a coincidence, then we might expect to find that the events making up this regularity each have alternative explanations in terms of a variety of other laws (i.e. each one falls under some other regularity). If this is the case, then the systematic account would not count this regularity as a law. For it would add little in the way of strength, since the events that fall under it also fall under other systematic regularities, while adding it would detract from the simplicity of the overall system.

This raises a difficult issue for the systematic theorist, and indeed for minimalists more generally. Might not this way of excluding coincidental regularities get it wrong and exclude genuine laws while including accidents? It might be that there is a large number of laws each of which has only a small number of instances. These laws may throw up some extraordinary coincidences covering a large number of events. By the systematic account the coincidence would be counted as a law because it contributes significantly to the strength of the system, while at the same time the real laws are excluded because their work in systematizing the facts is made redundant by the accidental regularities. Figure 1.2 shows a world similar to the one discussed

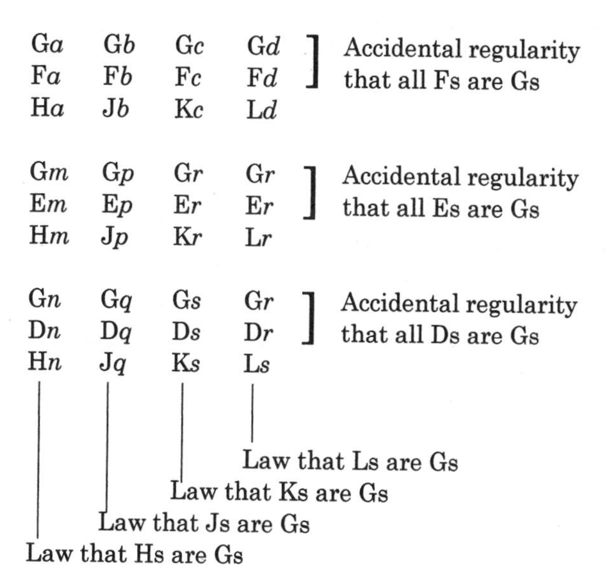

Figure 1.2

a few pages back. It has the same laws, but the additional facts mean that the accidental regularities now predominate over the genuine laws – the optimal systematization would include the regularity that Fs are Gs but exclude the law that Hs are Gs. If such a world is possible, and I think our intuitions suggest it is, then the systematic regularity theory gets it wrong about which the laws are. In this way the systematic theory fails to be materially adequate.

Laws, regularities, and explanation

I have mentioned that there are various things we want laws to do, which they are not up to doing if they are mere regularities. We have already looked at a related issue. One thing laws are supposed to do is support counterfactuals. Some opponents of the regularity theory think that this is sufficient to show that minimalism is mistaken, but I think that this is a difficult line to pursue (even if ultimately correct). More promising is to argue that laws cannot do the job of *explaining their instances* if they are merely regularities. Another issue is whether we can make sense of induction if laws are just regularities.

The key to understanding why a mere regularity cannot explain its instances is the principle that something cannot explain itself. The notion of explanation is one we will explore in detail in the next chapter. However, for the time being, I hope it can be agreed that for any fact F, F does not explain F – the fact that it rained may explain the fact that the grass is damp, but it does not explain why it rained.

As a law explains its instances, it explains all of them and, therefore, it explains the fact that is the uniformity consisting of all of them. But a regularity cannot do this because, according to the above-stated principle, it cannot explain itself. To spell this out in more detail, let us imagine that there is a law that Fs are Gs. Let there in fact be only four Fs in the world, *a*, *b*, *c*, and *d*. So the generalization "all Fs are Gs" is equivalent to

(A) (F*a* & G*a*) & (F*b* & G*b*) & (F*c* & G*c*) & (F*d* & G*d*) & (nothing is F other than *a*, *b*, *c*, or *d*).

The conjunction of all the instances is

(B) (F*a* & G*a*) & (F*b* & G*b*) & (F*c* & G*c*) & (F*d* & G*d*)

(A) and (B) are identical except for the last part of (A), which says that nothing is F other than *a*, *b*, *c*, or *d*. Let us call this bit (C)

(C) nothing is F other than *a*,*b*, *c*, or *d*

So (A) = (B) & (C).

Thus, to say (A) explains (B) is the same as saying (B) & (C) explains (B). But how can that be? For clearly (B) does not explain (B), as already discussed. So if (B) & (C) explains (B) it is by virtue of the fact that (C). However, (C) says that nothing is F other than *a*,*b*, *c*, or *d*. That other things are not F cannot contribute to explaining why these things are G. Therefore, nothing referred to in (A) can explain why it is the case that (B).

That we have considered a law with just four instances is immaterial. The argument can be extended to any finite number of instances. Many laws do have only a finite number of actual instances (biological laws for example). Nor does moving to an infinite number of instances change the nature of the argument. Even if it were thought that the infinite case might be different, then we could give essentially the same argument but, instead of considering the conjunction of instances, we could focus on just one instance. Why is *a*, which is an F, also a G? Could being told that all Fs are G explain this? To say that all Fs are Gs is to say if *a* is an F then *a* is a G, and all other Fs are Gs. The question of why *a*, which is an F, is also a G is not explained by saying that if *a* is an F then *a* is a G, because that is what we want explained. On the other hand, facts about other Fs, even all of them, simply do not impinge on this F.

It should be noted that this objection covers, in effect, all the forms of the regularity theory, not just the simple regularity theory. For more sophisticated versions operate by paring down the range of regularities eligible as laws and exclude those that fail some sort of test. But they still maintain that the essence of lawhood is to be a regularity, and the relation of explanation between law and instance is still, according to the regularity theorist, the relation between a regularity and the instance of the regularity. However sophisticated a regularity theory is, it cannot then escape this criticism. For instance, the deductive integratibility required by Lewis and Ramsey does not serve to provide any more unity to a law than is provided by a generalization. That the generalization is an axiom or consequence of the optimal axiomatic system does nothing to change the fact that it or the regularity it describes cannot explain its instances.

A case that seems to go against this may in fact be seen to prove to rule. I have an electric toaster that seems to have a fault. I take it back to the shop where I bought it, where I am told "They all do that, sir". This seems to explain my problem. Whether or not I am happy, at least I have had it explained why my toaster behaves the way it does. However, I think that this is an illusion. Being told that Emily's toaster does this, and Ned's and Ian's too does not really explain why my toaster does this. After all, if Emily wants to know why her toaster behaves that way, she is going to be told that Ned's does that too and Ian's and so does mine. So part of the explanation of why mine does this is that Emily's does and part of the explanation of why Emily's does this is that mine does. This looks slightly circular, and when we consider everyone asking why their toaster does this strange thing, we can see that the explanations we are all being given are completely circular.

So why does being told "They all do that" look like an explanation? The answer is that, although it is not itself an explanation, it *points* to an explanation. "They all do that" rules out as unlikely the possibilities that it is the way I have mistreated the toaster, or that it is a one-off fault. In the context of all the other toasters behaving this way, the best explanation of why my toaster does so is that some feature or by-product of the design or manufacturing process causes the toaster to do this. This is a genuine explanation, and it is because this is clearly suggested by the shopkeeper's remark that the remark appears to be explanatory. What this serves to show is precisely that regularities are not explanations of their instances. What explains the instances is something that explains the regularity, although the fact of the regularity may provide evidence that suggests what that explanation is.

Laws, regularities, and induction

If regularities do not explain their instances, then a question is raised about inductive arguments from observed instances to generalizations. The critic of minimalism, whom I shall call the full-blooded theorist says that laws explain their instances and that inferring a law from the observation of its instances is a case of *inference to the best explanation* – e.g. we infer that there is a law of gravitation, because such a law is the best explanation of the observed behaviour of bodies (such as the orbits of the planets and the acceleration of falling objects). (Inference to the best explanation is discussed in Chapters 2 and 4). Because, as we have seen, the minimalist is unable to make sense of the idea of a law explaining its instances, the minimalist is also unable to employ this inference-to-the-best-explanation view of inductive inference.[14] For the minimalist, induction will in essence be a matter of finding observed regularities and extending them into the unobserved. So, while the minimalist's induction is of the form *all observed Fs are Gs therefore all Fs are Gs*, the full-blooded theorist's induction has an intermediate step: *all observed Fs are Gs, the best explanation of which is that there is a law that Fs are Gs, and therefore all Fs are Gs.*

Now recall the problem of spurious (accidental, contrived, single case, and trivial) regularities that faced the simple regularity theory. The systematic regularity theory solves this problem by showing that these do not play a part in our optimal systematization. Note that this solution does not depend on saying that these regularities are in themselves different from genuine laws – e.g. it does not say that the relationship between an accidental regularity and its instances is any different from the relationship between a law and its instances. What, according to the minimalist, distinguishes a spurious regularity from a law is only its relations with other regularities. What this means is that a minimalist's law possesses no more intrinsic unity than does a spurious regularity.

Recall the definition of "grue" (see p. 18). Now define "emerire" thus:

X is an emerire = *either* X is an emerald and observed before midnight on 31 December 2000
or X is a sapphire and not observed before midnight on 31 December 2000.

On the assumption that, due to the laws of nature, all emeralds are green and all sapphires are blue, it follows that all emerires are grue. The idea here is not of a false generalization (such as emeralds are grue), but of a mongrel true generalization formed by splicing two halves of distinct true generalizations together. Now consider someone who has observed many emeralds up until midday on 31 December 2000. If their past experience makes them think that all emeralds are green, then they will induce that an emerald first observed tomorrow will be green; but if they hit upon the contrived (but real) regularity that all emerires are grue, then they will believe that a sapphire first observed tomorrow will be blue. As both generalizations are true, neither will lead this person astray. The issue here is not like Hume's or Goodman's problems – how we know which of many possible generalizations is true. Instead we have two true generalizations, one of which we think is appropriate for induction and another which is not – even though it is true that the sapphire is blue, we cannot know this just by looking at emeralds. What makes the difference? Whatever it is, it will have to be something that forges a link between past emeralds and tomorrow's emerald, a link that is lacking between the past emeralds and tomorrow's sapphire.

I argued, a couple of paragraphs back, that in the minimalist's view there is no intrinsic difference between a spurious regularity and a law in terms of their relations with their instances. And so the minimalist is unable to give a satisfactory answer to the question: What makes the difference between inducing with the emerald law and inducing with the emerire regularity? Being intrinsically only a regularity the law does not supply the required link between past emeralds and future emeralds; any link it does provide is just the same as the one provided by the emerire regularity that links emeralds and sapphires. (Incidentally, the case may be even worse for the systematic regularity theorist, as it does seem as if the emerire regularity should turn out to be a derived law because it is derivable from two other laws.)

The full-blooded view appears to have the advantage here. For, if laws are distinct from the regularities that they explain, then we can say what makes the difference between the emerald law and the emerire regularity. In the former case we have something that explains why we have seen only green emeralds and so is relevant to the next emerald, while the emerire regularity has no explanatory power. It is the explanatory role of laws that provides unity to its instances – they are all explained by the same thing. The contrast between the minimalist and full-blooded views might be illustrated by this analogy: siblings may look very similar (the regularity), but the tie that binds them is not this, but rather their being born of the same parents, which explains the regularity of their similar appearance.

A full-blooded view – nomic necessitation

The conclusion reached is this. A regularity cannot explain its instances in the way a law of nature ought to. This rules out regularity theories of lawhood. The same view is achieved from the reverse perspective. We cannot infer a regularity from its instances unless there is something stronger than the regularity itself binding the instances together.

The task now is to spell out what has hitherto been little more than a metaphor, i.e. there is something that binds the instances of a law together which is more than their being instances of a regularity, and a law provides a unity not provided by a regularity. The suggestion briefly canvassed above is that we must consider the law that Fs are Gs not as a regularity but as some sort of relation between the properties or *universals* Fness and Gness.

The term *universal* refers to properties and relations that, unlike particular things, can apply to more than one object. A typical universal may belong to more than one thing, at different places but at the same time, thus greenness is a property that many different things may have, possessing it simultaneously in many places. A *first-order universal* is a property of or relation among particular things; so greenness is a first-order universal. Other first-order universals would be, for example: being (made of) magnesium, being a member of the species *Homo sapiens*, being combustible in air, and having a mass of 10 kg. A *second-order universal* is a property of or relation among first-order universals. The property of being a property of emeralds is a second-order universal since it is a property of a property of things. The first-order universal greenness has thus the second-order property being a colour. *Being generous* is a first-order property that people can have. *Being a desirable trait* is a property that properties can have – for instance, being generous has the property of being desirable; so the latter is a second-order universal.

Consider the law that magnesium is combustible in air. According to the full-blooded view this law is a relation between the properties of being magnesium and being combustible in air. This is a relation of *natural necessity*. It is not just that whenever the one property is instantiated the other is instantiated, which would be the regularity view. Rather, necessitation is supposed to be stronger. The presence of the one property brings about the presence of the other. Necessitation is therefore a property (more accurately a relation) of properties. This is illustrated in Figure 1.3, where the three levels of particular things, properties of those things (first order universals) and relations among properties (second order universals) are shown. The arrow, which represents the law that emeralds are green, is to be interpreted as the relation of necessitation holding between the property of being an emerald and the property of being green.

The advantages of this view are that it bypasses many of the problems facing the minimalist. Many of those problems involved accidental regularities or spurious "cooked-up" regularities. What were problems for the minimalist now simply disappear –

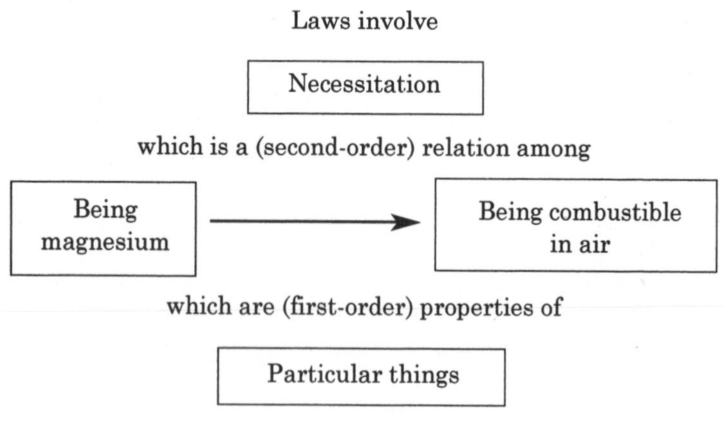

Figure 1.3

necessitation among universals is quite a different matter from a regularity among things. The existence of a regularity is a much weaker fact than necessitation between universals. Two universals can coexist in precisely the same objects without one necessitating the other. We would have something like the diagram in Figure 1.3, but without the top layer and so without the arrow. This is the case in a purely accidental regularity. In an accidental regularity every particular has both of two universals, but the universals are not themselves related. This explains why not every regularity is a law. It also explains why every deterministic law entails a regularity. If there is a law that Fness necessitates Gness then every F will be G. This is because if *x* is F then the presence of Fness in *x* will bring about Gness in *x*, i.e. *x* will be G. So the existence of a deterministic law will bring about the corresponding regularity without the reverse holding.

This deals with accidental regularities. Some of the more spurious regularities, for instance those employing grue-type predicates, are even more easily dealt with, because the cooked-up predicates do not correspond to universals. There need be no genuine property of being grue, and hence none to be related by nomic necessitation.

Nomic necessitation among universals also makes more sense of explanation and induction. Regularities, we saw, cannot explain their instances in the way that laws are supposed to, as to do so would be circular. No such problem arises for the full-blooded view. The fact of *a*'s being both F and G and the general fact of all Fs being Gs are both quite distinct from the fact of Fness necessitating Gness. The former are facts about individuals, while the latter is a fact about universals. And, as explained, while the latter fact entails the regularity that all Fs are Gs, it also goes beyond it, as it is not entailed by it. So reference to necessitation among universals as an explanation of particular facts or regularities is genuinely informative.

It also provides the unity among instances of a law that is necessary for induction to be possible. The thought is that if we see various different pieces of magnesium burn in air, we can surmise that the one property, being magnesium, necessitates the other, combustibility. If this conclusion is correct, then this relation will show itself in other pieces of magnesium. We induce not merely a resemblance of unobserved cases to those we have seen, but rather we induce a single fact, the relation of these two properties, which brings about the resemblance of unobserved to observed cases. In this way we should think of induction to facts about the future or other unobserved facts not as a one-stage process:

All observed Fs are Gs
∴ all unobserved Fs are Gs

but instead as a two-stage process

All observed Fs are Gs
∴ Fness necessitates Gness
∴ all unobserved Fs are Gs

What is necessitation?

So far so good. The idea of necessitation between universals seems to do the job required of it better than the minimalist's regularities, be they systematic or otherwise. The objection the minimalist has the right to raise is this. At least we know what we mean by *regularity*. But what do we mean by *necessitation*? It is not something we can see. Even if in a God-like manner we could see all occurrences of everything throughout time and space, we would still not see any necessitation. The necessitation of Gness by Fness *looks* just like the regularity of Fs being Gs. This may not be enough to show that there is nothing more to a law than being a regularity. But it does put the onus on the full-blooded theorist. For without a satisfactory answer the suspicion may remain that perhaps "Fness necessitates Gness" just means "there is a law that Fs are Gs", in which case the full-blooded theorist will really be saying nothing at all.

What then can we say about this notion of "necessitation"? A leading full-blooded theorist, David Armstrong, lists the following properties of necessitation, which you will recognize from our discussion:

(1) If Fness necessitates Gness then this entails that everything which is F is also G.

(2) The reverse entailment does not follow. Instances of Fness may only coincidentally also be instances of Gness, without there being any necessitation.
(3) Since necessitation is a relation, it is a universal. Furthermore, since necessitation is a relation among universals, it is a second-order universal.
(4) Since necessitation is a universal it has instances. Its instances are cases of, for example, *a*'s being G because *a* is F (*a*'s being F necessitates *a*'s being G).

Is this enough to tell us what necessitation is? It might be thought that if we listed enough of the properties of necessitation, then that would be sufficient to isolate the relation we are talking about. If I told you that I was thinking about a relation R, and that R is a relation among people that is used to explain why people have the genetic properties they do, then you would be able to work out that what I had in mind was something like the relation "being an ancestor of". So perhaps the properties (1)–(4) are sufficient to isolate the relation of necessitation.

In fact, it turns out that these conditions do not succeed in distinguishing the nomic necessitation view from the regularity view of laws.[15] To see this let the relation RL be taken to hold between universals F and G precisely when it is a law that Fs are Gs, according to the Ramsey–Lewis systematic account. The requirements on RL are as follows:

The properties F and G are RL related precisely when:

(a) all Fs are Gs;
(b) the above is an axiom or theorem of that axiomatic system which captures the complete history of the universe and is the maximal combination of strength and simplicity.

The point is that the RL relation does everything that necessitation is supposed to do according to Armstrong's properties (1)–(4). Let us take (1)–(4) in turn: (1) If F and G are RL related, then, by (a), all Fs are also Gs. (2) Because of (b) the reverse entailment does not follow. (3) The RL relation is a relation among properties. Hence it is a second-order relation. (4) We can regard "*a*'s being G because *a* is F" as an instance of the RL relation – when

F and G are RL related, *a* is both F and G, and the capturing of the fact that *a* is G by the regularity that all Fs are Gs contributes to the systematization mentioned in (b). (I discuss this point at greater length in the next chapter.)

As the requirements (1)–(4) can be satisfied by taking laws to be a certain species of regularity, these requirements cannot gives us any insight into necessitation that accounts for the most important metaphysical features of laws, i.e. those which we discussed above:

(i) a law explains its instances;
(ii) particular facts can count as evidence for there being a law;
(iii) it is possible for systematic regularities to diverge from the laws that there are (i.e. there can be a lot of systematic regularity for which there are no corresponding laws).

We could add (i)–(iii) to (1)–(4), which would then exclude RL relations. But to do so would be to give up on trying to give an illuminating explication of the concept of necessitation. There would be no reason for someone who is doubtful about the idea of necessitation to think that such a relation really exists. Thus an alternative account of necessitation that satisfies (1)–(4) and (i)–(iii) is still required. This is what I will try to provide next.

The criterial account of laws

It seems that neither minimalism or the full-blooded approach have succeeded in telling us what laws are. There is something they do have in common, which I suspect may be the reason for their failure. In both cases we have been offered a conceptual equation between laws and something else – either "it is a law that Fs are Gs = all Fs are Gs and such-and-such" or "it is a law that Fs are Gs = Fness necessitates Gness". Either way any talk of laws can be replaced by talk of something else, be it regularities or necessitation. The notion of explaining a concept by providing an equivalent for it in other terms is very popular in philosophy – giving an analysis of a concept by means of necessary and sufficient conditions. It is clear that not all concepts can be explained this way. If they could be, analysis would go on for ever, which it cannot, there being only a finite number of concepts in the English language, or it would be circular and so self-defeating. One alternative method of explaining a concept is by *ostensive* definition. This explains a term by pointing to the thing it names or a sample of the kind of thing named or an instance of the property named. So one might let someone know what you meant by "Gertie", "willow warbler", and "ultramarine" by pointing to Gertie, to a willow warbler, and something coloured ultramarine, respectively. Many philosophers have thought that a combination of these two kinds of conceptual explanation would suffice for all expressions of a language (except perhaps logical expressions). So ostensive definition would provide the basic material, and definition in terms of necessary and sufficient conditions would build up the rest from this initial stock.

This view, which has its roots in the ideas of Locke and Hume, has been most strongly associated in this century with Russell, Wittgenstein, and the philosophers of the Vienna circle, the Logical Positivists. We shall encounter another manifestation of it in Chapter 4 on realism. However, it has been largely discredited, and Wittgenstein himself in his later philosophy was keen to refute it. Nonetheless, the thought that, optimally, conceptual investigations should yield an analysis in terms of necessary and sufficient conditions has continued to exert its power over philosophers who would reject the picture of language on which it is based. For instance, full-blooded theorists would certainly oppose the thought that conceptual analysis would lead to definitions solely in terms which themselves could be defined ostensively; nonetheless, they may be happy to sign up to the idea that to explain a concept is to state the conditions under which sentences using it are true (what are called the *truth-conditions* of a sentence).

I propose that we take seriously the idea that there are other ways of explaining concepts. The one that I think is relevant to explaining the notion of law of nature employs the idea of a *criterion*. The role of criteria is summarized as follows:

(a) The relation between the criterion itself and what it is a criterion for is *evidential* (i.e. is not a matter of being a logically necessary or sufficient condition).

(b) The fact referred to in (a), that the relation is evidential, is a logical fact. That is, it is part of the *meaning* of the concept being explained or introduced that the criterion counts as good evidence for the concept's being correctly applied. The relation is not a contingent relation.

Let us see how this works. Let us look first at two cases that display one but not the other of the two roles. Consider a liquid that turns litmus paper blue, which is evidence that the liquid is alkaline. But it is no part of the *meaning* of "alkaline", that things which are alkaline turn litmus paper blue. So although the relation between blue litmus paper and the liquid's being alkaline is evidential (as in (a)), that relation has nothing to do with the logic or meaning of "alkaline", rather it is something we learn by experience. By contrast, take the relation between something being a female fox and it being a vixen. This is a logical relation – "vixen" just means "female fox" (as in (b)). But in this case the relation is not one of evidence. Once one knows that what one has is a fox and that it is female one knows all there is to knowing whether or not it is a vixen. One could not ask for additional evidence so as to make stronger the claim that one had a vixen, nor could one worry that additional evidence might come forward which would show that, although the creature was a female fox, it was not after all a vixen.

Some concepts share some of the features of both the alkali and the vixen cases. They share the evidential nature of the former and the logical/semantic nature of the latter. Imagine you saw someone sit down and tuck into a meal with gusto. You might conclude that they were hungry, or that they particularly liked the dish before them, or perhaps both. Why might you conclude that? Is it because you have learned from experience that hungry people are keen to eat or that people who like a dish eat with enthusiasm? Surely not, for it is part of what we *mean* by "hungry" that being keen to eat is a sign of hunger, and it is a matter of *logic* that if someone does something without outside encouragement and bearing a smile or satisfied demeanour we are entitled to surmise that they enjoy doing this thing. So the relationship between the behaviour and the state of mind we ascribe is in this case something to do with the meanings of the expressions we use. At the same time the relationship is not one of necessary or sufficient conditions. It is still evidential. For the evidence we have might be countered by other evidence. We might discover that the meal being eaten had been prepared by someone whom the diner wanted to please and not to disappoint, and so was giving the impression of enjoyment, despite not being hungry at all and not liking the meal. So the connection between enthusiastic behaviour and hunger or pleasure, even though it is a conceptual connection, does not constitute a sufficient condition. The fact that criteria may be countered by additional evidence is summed up by saying that criteria are *defeasible*. (In this case, the criteria are not only defeasible, and so not sufficient conditions, but they do not comprise necessary conditions either – for instance a hungry man might eat slowly and reticently in unfamiliar company.)

The proposal I want to make in the debate about laws of nature is that the relation between a regularity and a law is of the sort just described. More precisely, particular instances are criterial evidence for a law and the regularity is a limiting case of such evidence. This view differs from both of those we have so far been considering. On the one hand, in contradiction to the view of the regularity theorist, the relation between a regularity and a law is not one of identity, even in special cases, i.e. it is not the case that Law = Regularity + X. Nor, in contradiction to the view of the full-blooded theorist, is the relation a theoretical one. Rather, the regularity is logically good evidence, but potentially defeasible evidence, for the corresponding law.

Imagine that with a God's-eye view of the history of the world we detect a regularity to the effect that all Fs are Gs. Other things being equal this is evidence that it is a law that Fs are Gs. That it is so needs no explanation. But it is not *contingently* good evidence. It is not that we regard our belief in the regularity as justifying a belief in the corresponding law, because we have observed a correlation between regularities of the form *all Fs are Gs* and laws of the form *it is a law that Fs are Gs*, or because there is some other theoretical reason for thinking that such regularities and laws are linked. No, the relation is logical. It is part of the concept of a law that instances of the regularity that all Fs are Gs counts as evidence for its being a law that Fs are Gs.

The regularity theorist is right in this at least, that there is this conceptual connection between regularities and law. But, as remarked earlier, the relation is not one of identity – they are not the same thing (any more than eating with enthusiasm and hunger are the same thing). Mere identity, the position of the naive regularity theorist, is ruled out by straightforward defeasibility of the criterial conditions. We have already seen this in action. We may have a regularity, but we see that it is accidental – all Fs may be Gs, but for each of these things there is an explanation of its being G that has nothing to do with its being F. The grounds that one may cite for regarding a regularity as accidental are precisely the grounds for defeating the criterial evidence supplied by that regularity.

We can now see what the relevance of the systematization according to Ramsey and Lewis is. For systematization aims to rule out, as far as is possible, such defeating grounds. Therefore, if a regularity is subsumed under a RL system either as an axiom or as a theorem, then if (from a God's-eye view) we knew this regularity we would have the best possible grounds for claiming that there is a law corresponding to this regularity.

One issue that confronts the systematic regularity theorist is how to react to the possibility that there may be distinct systems that are equally optimal. Laws cannot be those regularities that appear in any of the equally optimal systems, for then conflicting counterfactuals would be supported by laws from different systems. Lewis' proposal is that a law is a regularity occurring as a theorem in all the optimal systems. This is unsatisfactory in that it leaves open the possibility that, even though the world may be a very regular place, the existence of widely different ways of systematizing the world might mean that the proportion of the world actually governed by laws, as defined by Lewis, is very small.

However, in the criterial view this problem does not arise. There cannot be conflicting laws, so the minimalist is forced to choose between competing regularities. However, there can be conflicting evidence. We might have the best possible evidence that P and the best possible evidence that Q, but there may be no grounds to choose between them, even when P and Q are incompatible with one another. This is all the criterialist need say about competing optimal regularities.

The criterial view inherits the best of the systematic regularity view, without taking on its problems. How does it fare in comparison with the full-blooded nomic necessity view? The problem here was that it was mysterious what the relation of necessitation between universals might be. On the one hand, the formal requirements on necessitation seemed to be satisfied by regularities of the systematic sort. On the other hand, Armstrong has rightly insisted, as I have, that laws go beyond any sort of regularity, however systematic. To speak metaphorically, they stand behind the corresponding regularities. We have seen that metaphor is to be spelt out in metaphysical terms:

(a) Laws explain their instances, regularities do not. (In particular laws explain the regularities themselves, while a regularity cannot explain itself.)
(b) Particular facts can count as evidence for there being a law.
(c) It is possible for systematic regularities to diverge from the laws that there are.

These insights cán be accounted for using the criterial view. Let us take first the possible divergence between laws and regularities, (c). When explaining the notion of a criterion, I made it clear that the criteria appropriate for a certain fact F could all be fulfilled, yet it not be the case that F. It is possible for someone to be eating with every appearance of pleasure yet not enjoy the food or be feeling hungry. Similarly, it is possible in the criterial view for there to be a lot of systematic regularity yet no corresponding laws. All the evidence would be in favour of the thought that there are such laws, and so it would be unjustifiable to assert that there actually are no such laws – but the assertion is not self-contradictory.

Since it is in the nature of criteria to be evidence, the evidential requirement (b) is satisfied immediately.

The fact that the existence of the criteria for a law and the existence of the law itself are two different things can be described as an *ontological gap*.[16] The ontological gap leaves room for the law to provide an explanation of its instances, as required by (a). The reason why a regularity cannot explain its instances is just that a regularity *is* its instances. The ontological gap shows that this is not so.

The full-blooded theorist uses the notion of nomic necessity to express the existence of this ontological gap. But as we have seen the attempt to treat nomic necessity as a theoretical idea does not work. We could ignore talk of nomic necessity altogether, now that the notion of lawhood has been explicated. However, I think it does serve a useful function. First, it reminds us that laws are not regularities. Secondly, it emphasizes the fact that laws are relations among properties. Thirdly, it shows that particular facts are of a different kind from laws and so may be explained by them. Therefore we may retain the full-blooded notion of nomic necessity in our expression of laws; what we now know is how this notion is to be explained, which is in terms of criteria.

Further reading

David Lewis presents the Ramsey–Lewis view in *Counterfactuals*. David Armstrong's *What is a law of nature?* gives readable but detailed accounts of all the main contenders in the debate on laws of nature, promoting his own necessitarian view. The latter is also promoted by Fred Dretske ("Laws of nature") and Michael Tooley ("The nature of laws"). Bas van Fraassen (*Laws and symmetry*) also discusses the various accounts of laws, preferring none of them (Chs. 1–5). He also promotes a no-law view (Ch. 8).

Chapter 2

Explanation

Kinds of explanation

One of the most common tasks of science is to provide explanations. The problem of induction addresses the question of prediction – saying what will happen inferred from what has been observed. But science is interested not only in questions of what will happen but also in questions of why what has happened did happen. If faced with an epidemic with new symptoms we may want science to tell us how the epidemic will progress, how many people will develop the symptoms, for how long, and in what concentrations. We will also hope that science will come up with an answer as to why this epidemic has broken out and why it is that people show the symptoms they do. Though distinct, the two questions are linked. One way of predicting how the epidemic will progress would be to discover and learn about the organism that causes it. That is, armed with the explanation of what we have already observed we may be able to predict what we will observe. Conversely, if there are alternative possible explanations available, then we may be able to discover the real explanation by checking what actually happens against the different predictions made on the basis of each of the competing explanations.

The varieties of explanation are numerous, though how many is a question of philosophical debate. Consider the following examples of explanations that we might come across:

(a) The window broke because the stone was thrown at it.
(b) The lump of potassium dissolved because it is a law of nature that potassium reacts with water to form a soluble hydroxide.
(c) He stayed in the café all day in the hope of seeing her again.
(d) His dislike of gerbils stemmed from a repression of his childhood fear of his father.
(e) Cheetahs can run at high speeds because of the selective advantage this gives them in catching their prey.
(f) Blood circulates in order to supply the various parts of the body with oxygen and nutrients.

Explanations may be classified. Those just given illustrate the following kinds of explanation:

(a) causal explanation;
(b) nomic explanation (explanations in terms of laws of nature);
(c) psychological explanation;
(d) psychoanalytic explanation;
(e) "Darwinian" explanation; and
(f) functional explanation.

As remarked earlier, it is an interesting philosophical question whether these are really all distinct kinds of explanation, indeed whether these and any other kind of explanation there may be are in fact facets of one all-encompassing kind of explanation. There is a clear sense in which, as they stand, not all these explanations are the same. For instance, the causal explanation is not a nomic explanation. If it were, there would have to be a law that stones thrown at windows always break them. But they do not. A small stone thrown weedily at a toughened plastic window will not break it. At the same time, the nomic explanation is not a causal one. The chemistry of potassium does not cause it to react. The laws of chemistry, we have seen, are relations between, in this case, potassium in general and water in general; they hold for all pieces of potassium, whether in contact with water or not. What causes one piece to react rather than another was that someone put the first piece into a beaker of water but not the second. Whether psychological and psychoanalytic explanations are causal are questions in the philosophy of mind, which are better discussed elsewhere. But it is clear that they are not nomic explanations, since they do not have the generality characteristic of natural law. There is no law-like relation between spending a long time in cafés and waiting to see people. What I have called "Darwinian" explanations are interesting because they form the basis of much explanation in biology but do not seem to be either causal or nomic, the sort that are most common in physics or chemistry. Causal and nomic explanations are effective at explaining both general and particular cases, while Darwinian explanations are good only for the general case. The selective advantage explains why cheetahs as a species can run fast – it explains why a certain trait is widespread rather than rare in a population. And so it does not explain why this particular cheetah has the trait of running fast (which it might have had whether or not the trait is widespread). That has something to do with the precise nature of its physiology, which is a causal or nomic affair.

So the non-mental forms of explanation at least are all different in that they cannot be readily transformed from one into another. "A because B" may be a nomic explanation without being a causal or Darwinian one, and so on. An important feature of an explanation focuses on the sort of information that it gives. This in turn will typically be related to the interests of those offering or being offered the explanation. So a gardener faced with a broken glasshouse pane will want to know the cause, the chemical engineer wanting to devise a means of producing a certain chemical will want to know the laws governing the various reactions, and the plant breeder looking for a high-yield variety may be more interested in Darwinian explanations. The gardener with the broken pane does not want to know the laws governing the fractibility of glass, nor may the plant breeder want to know about the physiology of a high-yield plant, since this may be something outside his or her control. Nor does the chemical engineer learn much if you tell him or her that the beaker exploded because I dropped some of the stuff from the brown jar into it. That may be the cause, but that is not very helpful in learning the general features of things unless the chemist is also told the conditions under which this occurred (what were the temperature and pressure, were there any unusual gases around, and what was in the beaker anyway?).

This seems to put a very subjective slant on explanation. Indeed, some philosophers have thought that explanation is hopelessly subjective, so subjective even that it should have no part in proper science. As I shall explain, I do not think this is right. Indeed, it had better not be right because explanation is closely related to inference; the standard form of scientific inference is Inference to the Best Explanation. Inference to the Best Explanation says, for instance, that when seeking the cause of a phenomenon we should look for that possible cause that best explains the phenomenon. If explanation is subjective, then that which our best explanation is will also be a subjective matter. If that is right, our inferential practices and hence the strength of our claims to scientific knowledge will also be subjective. I hope to show that science is on a better footing than that. Consequently, it had better be that explanation is not subjective.

Thankfully, it is an error to conclude, from the fact that the usefulness of different explanations is relative to one's interests, that explanations are subjective. The value of information generally is interest relative. The use of information about how to get to Edinburgh will depend on where you are coming from, whether you have a car or are using public transport, how important time and cost are to you, etc. But this does not undermine the objectivity of the information. It is not a subjective matter that you can get from York to Edinburgh by train and that to get there from Shetland you need to take a boat or plane. Better than calling this feature of explanation subjective is to call it *epistemic*. This emphasizes the fact that one way in that explanations differ is in the sort of information they provide.

If I am right then there is also an "objective", non-epistemic, component to explanations. This component consists of the things that must exist for A to be able to explain B, and the relations those things have to one another. For example, someone tells you that the existence of tides is explained by the Moon's gravitational pull. What makes this true? For one thing, the Moon must exert a gravitational pull – if it did not there would be no such thing as "the Moon's gravitational pull" to explain anything. Similarly, tides must exist for them to be explained. There must be something else as well. The Moon's gravity and tides might both exist without one explaining the other. (Jupiter's gravity and the continental drift on Earth both exist without the former explaining the latter.) So what else must be true for there to be an explanatory relation between two things?

What this might be we shall see in a moment – first we must give this feature of explanation a name. Since we have replaced "subjective" by "epistemic" to characterize the other aspect of explanation, it is right that we should avoid "objective" for this aspect. For want of a better expression, I propose to call this the *metaphysical* component of explanation.

Although the epistemic features of different explanations may differ, it may be that their metaphysical content is more unified. Whether or not this is the case has important consequences for our view of both science and other aspects of the world. For instance, some philosophers think that psychological explanations have the same metaphysical component as nomic explanations and causal explanations. That is, although psychological explanations do not mention any laws, it is claimed that certain sorts of law (laws of the brain's physiology) must exist for a psychological explanation to be true. If this is right, and if the relevant laws or causes are deterministic, then so are the corresponding psychological relations. This has implications for the debate on free will. In the case of the non-life sciences the explanations we are interested in are causal and nomic. If the metaphysics of these two is fundamentally the same, then the same sort of fact underlies the truth of a causal explanation as underlies a nomic explanation. This is the view I take; in particular, I think that laws are fundamental and that where there is a cause there is always a set of laws that encompass the cause. If I am right then identifying a cause will never be the final word in a scientific investigation, but will always be open to supplementation by finding the laws underlying the cause. But if I am wrong then there may be causes that are not subject to laws. So some things may happen because they are subject to laws and others because they are the effects of causes, and these are just two different ways in which things may come about. Thus even if we, *per impossibile*, had discovered all the laws of nature, there would, according to this view (with which I disagree), still remain the business of discovering the causal relations that there are.

Later we will see how it is that causes are bound up with laws. For the moment we will stay with the problem of characterizing explanation. One clarification needs to be made before we proceed further. We use the word "explains" in more than one way. We talk of *people* explaining things and of *facts* explaining things. We might say that the gravitational force exerted by the heat and radiation energy of the Sun explains the observed perihelion of Mercury (by virtue of the relativistic equivalence of energy and mass). Here a fact explains another fact. But people may also explain things, as when one might say "Einstein explained the perihelion of Mercury with his theory of relativity".

When we talk of facts explaining, our notion of explanation is *factive*, that is to say that the fact in question must exist and it must explain the thing being explained. The gravitational force of the Sun explains the perihelion of Mercury only if, first, there is such a gravitational force and, secondly, if it really does account for the perihelion of Mercury.

When we talk of people explaining, "explanation" is typically being used *non-factively*. We might say "Some scientists explained the perihelion of Mercury by postulating an inner planet, which they called Vulcan". The interesting thing here is that we say this even though we know there is no such inner planet. In this sense the explanation someone gives may be completely false.

It may be that thinking about the non-factive sense also exacerbates the feeling that explanation must be subjective. But if we keep the two senses apart we can avoid this error. When we say that Priestley explained (non-factively) combustion as the removal of phlogiston from the air, we mean that Priestley believed or *proposed* that the (factive) explanation of combustion is that phlogiston is removed from the air. Since non-factive explanation is a matter of what people take to be the factive explanation, we will in what follows be concerned primarily with factive explanation, though non-factive explanation has its part to play too.

Hempel's models of explanation

Having made the appropriate distinctions between factive and non-factive explanation and between their epistemic and metaphysical components, let us look at how philosophers have attempted to characterize what we may now call the metaphysical component of factive explanation. The philosopher whose work we will consider first is Carl Hempel, who was born in Germany, but like many philosophers left under threat of Nazi persecution. He thereafter worked in the USA.

Hempel proposed that to explain a fact is to show how that fact could be subsumed under a law or several laws together with various antecedent conditions. We will soon see how this works. This approach to explanation is called the *covering-law* view, since explanation is a matter of stating a law that covers the case in question. Since the focus is on laws, this is an account of nomic explanation. We shall see how causal explanation is also captured by this account. There are two versions of this view: the *deductive–nomological model* of explanation, and the *probabilistic–statistical model* of explanation. What we mean by "model" is a set of conditions that together form a template, such that if something satisfies the conditions and so fits the template, it explains the fact to be explained. (The fact to be explained we call the *explanandum*, and that which does the explaining we call the *explanans*.)

We will look first at the deductive–nomological model. This model is so called because a phenomenon is explained if it can be deduced from the law plus certain auxiliary statements concerning the explanandum. Recall the lump of potassium that is seen to react, producing a gas, and disappear when placed in a beaker of water. Key to explaining the phenomenon is the chemical law that potassium reacts with water to produce a soluble hydroxide, as represented by the equation:

$$2K + 2H_2O \rightarrow 2K^+_{(aq)} + 2OH^-_{(aq)} + H_{2(gas)}$$

To be more specific, laws of reaction will specify the conditions, such as the temperature of the water and the vapour pressure, under which the reaction will take place and the degree to which these affect the rate of reaction and the degree of reversibility. Thus to use the law to explain what is observed we will need to state that the conditions mentioned in the law are met by the apparatus in question, i.e. that the temperature of the water and the vapour pressure were such that the rate and completeness of the reaction match what is observed. So what we have done is to specify a law or set of laws and a description of the circumstances of the observed phenomena such that we can deduce from a combination of these two that the observed reaction and dissolving is what should occur. In general, the form of the explanation is:

Law	$L_1, L_2, ..., L_n$
Conditions	$C_1, C_2, ..., C_m$
entail	———————
Explanandum	$O_1, O_2, ..., O_k$

where $L_1, L_2, ..., L_n$ are the various laws used in the explanation (there may be more than one), $C_1, C_2, ..., C_m$ are the circumstances or conditions of the observation, and $O_1, O_2, ..., O_k$ are the observed phenomena in question. (Hempel regards the entailment as holding between statements of laws, conditions, and explananda.)

A simpler case might be the following. A quantity of gas is contained in an air-tight syringe. The plunger is depressed to reduce the volume by a third. The pressure of the gas is measured beforehand and again afterwards. The second measurement shows that the pressure has increased by 50 per cent. The temperature of the room is the same at the time of the first measurement as the second. The law needed to explain this is Boyles law – i.e. under conditions of constant temperature, pressure and volume are inversely proportional. A description of the circumstances says that the temperature remained constant. These two plus the fact that the volume was decreased by a third entail the fact that the pressure increased by 50 per cent.

Law	Under conditions of constant temperature, $PV = k$
Conditions	The temperature remains constant
Conditions	The volume decreased by one-third
entail	————————
Explanandum	The pressure increased by 50%

It is because the observed phenomenon, the explanandum, can be deduced logically from a statement of the law and relevant conditions that we have a deductive–nomological explanation. That is, given the laws and the conditions we could expect the phenomenon in question to arise.

As we normally use the word, "expect" does not always mean expect with certainty; it can also mean to expect with a high degree of probability. Hempel takes this into account in his description of *probabilistic–statistical* explanation (also called *inductive–statistical* explanation). This is similar to deductive–nomological explanation, but instead of entailing the explanandum a probabilistic–statistical explanation makes it very likely. So the form is:

Law	p(O/F) is very high
Conditions	Fi
make likely	════════
Explanandum	Oi

Here the probability of O given F is very high. Under circumstances i we find that F. Hence, in these circumstances, it is very likely that O. This, according to Hempel, is sufficient to explain O. For example, one might explain the fact that it rained by reference to the confluence of a cold front and a bank of warm sea air. The cooling of sea air does not guarantee rain, but does make it likely. So in this case, given the explanation, the explanandum is very probable but not certain.

Not all the explanations we actually use appear to fit one or other of these models; indeed perhaps very few do. Hempel does not regard this as a refutation of his account of explanation. The role of this account is not to give descriptions of our actual explanatory practices, but to present an idealization of them. The case of Hempel's models of explanation is rather like that of formal logic. Nobody uses formal logic when debating or arguing or proving something. Formal logic therefore does not describe such practices, but it can be used to evaluate them. If a formal proof comes up with a conclusion contradicting the conclusion of an informal argument then one has a very good, perhaps the best, reason for thinking that the informal argument is fallacious. If we can formalize an informal argument then we will be more willing to rely on its validity. But in so doing we may well be losing something too. Formal arguments tend to be much longer and hence less easy to grasp. They will involve a lot of detail unnecessary to the purpose at hand.

Similarly a full deductive–nomological explanation has the virtue of being the best of its kind – it really does explain the explanandum rather than just seeming to. It provides a model to which "informal" explanations aspire. But at the same time it does have its disadvantages. There will be a lot of detail, particularly in specifying the conditions C_1, C_2, ..., C_m, but perhaps also among the laws L_1, L_2, ..., L_n, that will be trivial, obvious, or unnecessary for the purpose to which the explanation is being put. In everyday explanations this detail is usually left out. So in explaining the fact that the potassium reacted and dissolved in water one will refer to the law governing the reaction, but will quite likely leave out the detail involving rates of reaction or the fact that there was sufficient water in the beaker or that the room was at normal pressure, although one could add such detail for the sake of rigour. In the case of explaining rainfall there will be features of local geography that will affect the precise operation of the law, that one need not mention in order for the informal explanation to count as an explanation. But in both cases it must at least be possible, in principle, to fill in this detail. "In principle" here does not mean "possible for the person doing the explaining". It is a more objective notion than that. What is meant is that the relevant facts exist – that the unstated conditions are fulfilled and that there are laws of the relevant sort, even if the person providing the explanation cannot specify all these conditions and laws. So the actual explanations we give (if they are valid explanations) are, to use the jargon, *elliptical* – that is they are like perfect deductive–nomological or probabilistic explanations, except that some of the detail is omitted because of our ignorance or its not being relevant to the purpose of the explanation.

The fact that our everyday explanations are elliptical, incomplete, and condensed versions of full explanations allows us to subsume *causal* explanations under Hempel's models. Take our case of explaining the broken window by reference to the stone thrown at it. This was a causal explanation. If we were able to, we could specify all the relevant conditions surrounding the throwing of the stone and its breaking the window (the velocity and direction in which the stone was thrown, the mass and shape of the stone, the nature of the air, the thickness and constitution of the glass, and so on) and also the relevant laws (the laws of motion, laws governing air resistance, laws governing the behaviour of materials under stress, and so on). Once we had done this, then we would have specified all the laws and all the conditions necessary for a full deductive–nomological or probabilistic explanation, the conclusion of which is that the glass fractures. Our actual explanation leaves out all the laws and all the conditions except the throwing of the stone. Even that statement is less specific than the detailed description of the way in which the stone was thrown in the full explanation. Hence there is an epistemic difference between nomic and causal explanation, although metaphysically they are the same. They provide different sorts of information – the nomic explanation states an important law governing the phenomena, whereas the causal explanation cites a salient feature of the background conditions. But the form of the full explanation is the same for both.

An important feature of Hempel's covering law view is that it embodies a particular conception of the relation between explanation and prediction. I have already mentioned that, according to the covering law approach, to explain something is to show that it was to be expected. Given the relevant laws and conditions, the deductive–nomological model shows that the explanandum had to be, while the probabilistic–statistical model shows that it was very likely. In both cases the information furnished by a full explanation would have allowed us to predict the event – either with certainty or with a high probability of being right. This relation can be summed up as: *Explanation is prediction after the event and prediction is explanation before the event.*

Problems with the covering law approach

Compelling though it might be, the covering law approach has several flaws. Some of these require minor improvements to the model. Others, it seems, require more drastic treatment. Peter Achinstein gives the following Hempelian explanation:

Law	Everyone who eats a pound of arsenic dies within 24 hours
Conditions	Jones ate a pound of arsenic
entail	______________________________
Explanandum	Jones died within 24 hours

In Achinstein's example we are asked to imagine, quite reasonably, that the first is indeed a law, and also that it is true that Jones ate a pound of arsenic and that he died within 24 hours of so doing. But Achinstein goes on to tell us that what really killed Jones was his being hit by a bus. The point is that the deduction given meets all of Hempel's requirements, without being a real explanation. At first this seems to present a considerable difficulty for Hempel. We cannot tell that there is anything wrong with this proposed explanation of Jones' death just by examining the explanation itself. For it might well have been a satisfactory explanation of that occurrence. There will be other people who do indeed die of arsenic poisoning without their death being pre-empted by a bus, by falling down stairs, or some other more immediate manner of decease. So we cannot in general say that there is something wrong with an explanation like this, because sometimes we will want to use just such an explanation. It looks as if what prevents this from being the explanation in this case is that another and better one is available. In turn this suggests that Hempel's project of finding a *model* of explanation cannot succeed. The problem is that the fact that something fits the model is no guarantee that it is an explanation, for that depends on what other explanations are there to be found.[17]

Achinstein's case is important. It shows that the explanations we tend actually to employ are fallible in that, even if the premises of the proposed explanation are true, it may be that they do not explain the explanandum. David-Hillel Ruben[18] points out that this shows an asymmetry between explanation and prediction. Knowing that he had eaten a pound of arsenic would allow someone to predict his death, even if, as in Jones' case, it is not this that kills him. For good prediction we need only know facts that are appropriately correlated with the fact to be predicted. Correlation is not enough for explanation, since correlations can go both ways whereas explanation goes only one way. So prediction and explanation need not always be symmetrical. There are other cases that show asymmetry. It may be that A and B are perfectly correlated, but it is A that causes B, not vice versa. We could then predict A, given B, although B does not explain A. This raises a second problem with Hempel's model, since the latter does treat prediction and explanation symmetrically. Take the case of the gas in the syringe. We explained the increase in pressure with reference to Boyle's law and to the reduction in volume by depressing the plunger:

Law	Under conditions of constant temperature, $PV = k$
Condition	The temperature remained constant
Condition	The volume decreased by one-third
entail	______________________
Explanandum	The pressure increased by 50%

But, because the relation between pressure and volume is perfectly symmetrical as far as Boyle's law is concerned, the following is also a perfectly valid argument:

Law	Under conditions of constant temperature, $PV = k$
Condition	The temperature remains constant
Condition	The pressure increased by 50%
entail	______________________
Explanandum	The volume decreased by one-third

This second argument also fulfills the requirements of the deductive–nomological model. It follows that we are able to explain the decrease in volume by the increase in pressure. But of course that is the wrong way round. What caused the decrease in volume was my hand on the plunger, and the decrease in volume in turn caused the increase in pressure. There are many cases of the covering law model getting the direction of explanation wrong. Any case where the law in question is an equation of quantities will be subject to it. Other cases arise where the law states the coexistence of two properties that are the effects of a common cause or where an effect has a unique cause. These cases allow us to infer one effect from another or a cause from its effect. Clearly none of these inferences is an explanation. For instance, all animals with a liver also have heart. So we can deduce from this plus the existence of Fido's liver that he also has a heart. But his liver does not explain why he has a heart. Only bubonic plague has the symptoms of festering buboes. So we can deduce from this fact plus the presence of the symptoms that this patient has the disease. But the symptoms do not explain the disease. It is the other way around.

Some of these cases have something in common with Achinstein's poisoning case. Some of the "pseudo-explanations" that fit Hempel's model, but which we recognize as not being explanations, might have been genuine explanations in other circumstances. In the syringe case the change in volume caused the change in pressure, not the reverse. But take a balloon filled with air and submerge it in water. The deeper the balloon is taken the greater the pressure. The volume occupied by the balloon will consequently decrease. This is then a case where a change in pressure causes a change in volume. Again we cannot tell whether an explanation is an explanation just by looking at its components and their relation.

The third kind of problem for Hempel is best illustrated by the probabilistic–statistical version of the covering-law model. In this version the explanandum is not entailed by the law and conditions, but is made highly likely by them. This means that it is impossible to explain events the occurrence of which is very unlikely. A build up of static electricity might cause a fire even though that effect might be extremely unlucky and unlikely; if the static had not built up the fire would not have happened – it explains the fire even though the static build up does not make the fire probable or expected.

The problem more generally with Hempel's approach, as exemplified by this case, is to regard an explanation as that which makes the explanandum expected. For sure, there are many cases where we would expect an effect on the basis of seeing something that would cause it. But there are also many other cases where we have expectations for other reasons, e.g. expecting the existence of a cause on the basis of seeing its effect. This shows that the covering-law model is too generous, there are many reasons for expectation that clearly are not explanations. At the same time Hempel's approach cannot account for certain genuine explanations that do not give any expectation, because the chances of the phenomenon are so low. But even unlikely events can have explanations.

So the moral is that explanations cannot be equated with (law-based) arguments that give us reason to expect the explanandum. A more plausible and related thesis is that explanations *increase* our expectation of the explanandum. This is clearly not sufficient for an explanation, since it does nothing to address the problem of the direction of the explanation. But is it necessary that explanations should raise our expectation of the effect? The static electricity case suggests that this is a good proposal. For without the static we should have expected no fire at all. With the static the chances of a fire are still small, but they are greater, and correspondingly our expectation should increase.

There are, however, counter-examples even to this proposal. Someone is likely to die as a result of a tumour if they do not have an operation. The operation carries with it some small risk of life-threatening complications. So the patient has the operation and the tumour is successfully removed. But the complications, although unlikely, do occur and the patient dies from them. So the information that the patient underwent the operation both explains the patient's death while at the same time reducing the probability of death.

A numerical case that leads to the same conclusion is this. Consider virus V which is widespread, occurring in 50 per cent of the population, evenly spread among all groups. Symptoms S occur only as a result of the virus V. These symptoms are observed in 80 per cent of those people who are infected with V. But within a small minority of the population V never causes symptoms S except in the presence of bacterial infection B (which only ever occurs in this subpopulation). In the presence of B the chances of someone in this subpopulation developing the symptoms is 30 per cent. Our expectation that someone taken at random should display the symptoms is 50 per cent × 80 per cent = 40 per cent. Freddie shows symptoms and we discover the virus. Can we explain the symptoms S in Freddie's case? Yes, the explanation is that he has the virus. As stated, the expectation of V given S is 80 per cent. So knowing that Freddie has V raises the expectation of S from 40 per cent to 80 per cent. The model says explanations raise expectations, which is what we have here. So far so good for the model. Charlotte also shows symptoms S. We discover that she belongs to the subpopulation of people that never display S except in the presence of bacterium B, and sure enough she has that bacterial infection. So the full explanation of Charlotte showing S will involve reference to her being a member of this special subpopulation, to her having the virus V, and her being infected with B. This will be a satisfactory explanation. But our expectation that members of this subpopulation will display S, even when infected with both V and B, will be 30 per cent. So this explanation has actually reduced our expectation of the explanandum (remember that for the population as a whole the expectation of S is 40 per cent).

I think the notion of a probabilistic explanation needs careful examination. There are two sorts of case. The first is where the use of probabilistic explanation reflects our ignorance of a deterministic cause. Without the static electricity a fire could not have occurred, but even with the static the chances of a fire were low. Why are the chances low? It may be that for the static build up to lead to a fire a particular set of conditions must arise – the air must be damp enough for a discharge to occur but there must also be sufficient dry material around to catch light, and so on. This combination of circumstances rarely occurs, so the chances of a static build up causing a fire are low. However, in precisely the circumstances that did occur the fire was inevitable. If there is that amount of static electricity and a metallic object that distance away and if there is that level of moisture in the air, then there will be a discharge. If there is a discharge and that concentration of petrol fumes and exactly that material with that degree of dryness then it will catch fire, and so on. The point here is that, given the laws of nature and precisely those circumstances, there is a deterministic causal chain from the static build up to the fire. Being deterministic, each part of the chain has its own deductive–nomological explanation, so the explanation as a whole is a deductive–nomological explanation. Again, this is a case of an elliptical deductive–nomological explanation, one that does not include all the relevant detail that is possible. It has the form of a probabilistic explanation, because we may not be able to spell out some of this detail precisely – or we may have no need to. Instead, what we can report is that the sort of circumstances in question have such-and-such chance of occurring.

A different case is where the explanandum event is inherently probabilistic, for instance the decay of an atomic nucleus. Here the decay is not determined by the circumstances or the history of the nucleus. Two nuclei may be identical in every respect but one may decay while the other does not. Nonetheless, such events are subject to natural laws insofar as the laws determine the probability of decay. Different kinds of nuclei have different probabilities of decaying, a fact that is manifested in the different half-lives of various isotopes. The probabilities can be changed, e.g. by bombarding the nuclei with α particles. Such phenomena arise from the essentially probabilistic nature of the quantum laws governing the structure of the nuclei.

It is widely thought that the probabilistic–statistical model or something like it may be used to explain the occurrence of law-governed non-deterministic events such as nuclear decay. But this seems to me to be quite possibly a mistake, even in the best case for the model where the decay of a nucleus is explained by reference to a law that makes that decay highly likely during a given period

(e.g. a thorium-234 nucleus has a probability of 0.9 of decaying within 80 days.)

The reason for my doubt is that the supposed explanations of intrinsically probabilistic events fail to allow for *contrastive explanations* of those events. A contrastive explanation explains why one thing happened but not another. So the explanation of why Ralph rather than Matthew developed malaria is not that Ralph was bitten by an infected mosquito, as they were both bitten. The contrastive explanation is that Matthew has the sickle-cell trait (which confers some protection against malaria) while Ralph does not. Contrastive explanation seeks some difference in the causal history of the contrasting cases or a difference in the laws governing the two cases. Such differences cannot be found for otherwise identical nuclei one of which decays and one of which does not. Take two thorium-234 nuclei, one of which decays while the other does not. What explains why only one nucleus decayed? There is no explanation for this. There is, until the decay, no difference whatsoever between the nuclei. The most we can say is that one decayed and the other did not, and there is simply no explanation for this fact. That being so, it is difficult to see how the relevant law of nuclear decay explains the behaviour of one atom but not the other when both are equally subject to that law. The conclusion one should draw is that the law explains neither the occurrence of the decay nor its non-occurrence.

This conclusion has an air of paradox. How can a law of nuclear decay not explain nuclear decay? Is that not what laws are supposed to do? What else might the law explain if not nuclear decay? Well, there is an answer to that question, and it depends upon our accepting the notion of *objective chance*. The idea here is that each and every thorium-234 nucleus has the following property: its probability of decaying within 80 days is 0.9. This probability is not a statistical feature of large collections of thorium-234 nuclei. Rather it is an objective and intrinsic quality of each individual thorium nucleus. In the case of the two nuclei the law explains the fact that both atoms had a chance of decaying of 0.9; it does not explain the actual decay or failure to decay. These things simply are not to be explained – that is why they are indeterministic. But they can be expected or unexpected on the basis of law-governed chances. (For more on objective chance, see Chapter 6.)

This discussion of probabilistic explanation suggests two things:

(a) The idea that explanations raise expectations is unworkable; the concept of expectation is doing too much work here; it is an epistemic notion, while what we need is a more metaphysical notion.
(b) The idea that we can explain indeterministic events is misplaced, although what we can explain is the chance they have of occurring.

A holistic approach to explanation

Where does Hempel's covering-law model now stand? I think we should jettison the probabilistic version and confine our attention to the deductive–nomological model and explanations of deterministic events, properties, and facts (including objective chances). The deductive–nomological model itself suffers from two problems: Achinstein's problem of pre-emption, and the problem of the direction of explanation where the laws in question are symmetrical.

I suggested that the pre-emption problem and symmetry problem are similar, in that both suggest that what appeared to be explanations, and indeed might in different circumstances have been explanations, failed to be explanations. And we saw that they failed to be explanations because other explanations were available that did the work the purported explanations were supposed to do. In the Achinstein cases, A might have explained B, but does not because C in fact explains B. In the symmetry cases, A does not explain B because B explains A. The similarity to the former may be emphasized by considering that we see that B explains A, and not vice versa, because we know that C independently caused B. We know that the change in volume caused the change in pressure in the syringe and that it was not the change in pressure that caused the change in volume, because we know what caused the change in volume (my pressing on the plunger).

Some writers have taken this to show that we need to add something else to Hempel's model or even come up with an alternative model altogether. But either such attempts have failed or they have introduced notions such as "cause". For instance, one might say that A explains B if, and only if, A causes B. One advantage of Hempel's model is that causal explanation can be seen as a variety of nomic explanation. A causal explanation is a partial nomic explanation where what is given is one of the conditions. The condition is a cause or causal factor. As the covering-law model is, in general, faced with problems, so is the use of it to explicate causal explanation. But one might fear that to use the notion of cause is buying a solution too cheaply. The problems that beset the covering-law model are likely to recur if we try to explain the notion of cause. Recall the unfortunate Jones who was hit by a bus after ingesting arsenic. Which caused his death? An account of causation will have to deal with the same issues as arise for explanation in general. So I think that causal models of explanation do not really solve the problems we have looked at, but simply postpone them.

Secondly, not everything can be dealt with by a causal account. There are still genuinely nomic explanations that need to be accounted for. Newton's laws can explain Kepler's laws but do not cause them. There would still be a need for an account of this sort of explanation. So I think that it is worthwhile seeing whether some account of explanation might be retained in a form that would encompass both nomic and causal explanation.

I suggest that finding an answer to the problems of explanation does not lie in constructing alternative *models* of explanation. Rather, we should eschew the idea that explanation can be explained in terms of a model in favour of a holistic approach. The idea of a model of explanation is this: if something fits this model then it is an explanation. It is a template against which we can judge something to see whether it should count as an explanation. I think that it is this way of approaching the concept of explanation that is the source of the problems.

The holistic approach takes its cue from the fact that we know in the problem cases that the pseudo-explanations are mistaken because we know what the real explanations are. We know that the arsenic does not explain Jones' death because we know that the collision with the bus does. We know that the increase in pressure does not explain the reduction in volume because we know that my pushing the plunger does.

To determine what explains an event we have to consider not

just that event and the laws that cover that case, we also have to consider other possible explanations of that event and explanations of other related facts. Consider first the pre-emption case. Note that the law in question, that everyone who eats a pound of arsenic dies within 24 hours, is not a basic law, but a derived one. This law can be derived from three laws of a more basic nature: (a) arsenic prevents the functioning of enzymes; (b) failure of enzymes to function causes bodily organs to fail; and (c) acute organ failure brings about death. Each of these laws can be explained in terms of more basic laws. Even so, we can already see that Jones' death does not instantiate any of these laws. For it is not organ failure that leads to that event, but the collision with the bus (the latter severs his spinal column). The fact of his death can be explained by the law that severing one's spinal column leads to death, but not the law that organ failure leads to death (since this did not occur). Jones' death is not explained by his ingesting arsenic, since the law in question is derived from other laws, none of which can explain that death.

Turning now to the symmetry problem, the change in volume is explained not by the change in pressure but by my pressing the plunger. The point here is not that an event can only have one explanation – perhaps there is genuine causal overdetermination, where an event is caused by two causes, both of which would have been sufficient alone. There are other considerations. For instance, if the change in pressure were to explain the change in volume, what then explains the change in pressure? (I take it that this explanation is asymmetrical – if A explains B, B cannot explain A.) Furthermore, as in the case of Jones' death we can look at underlying explanations in terms of more basic laws. We can show how the change in volume leads to a change in pressure, using the kinetic theory of gases. However, in this case we cannot provide a deeper level explanation of the reverse.

These cases show that whether something is an explanation depends on its relations with other facts and explanations, including explanations at a more basic level. This means that the idea of a model of explanation is too simple – the template story cannot be the whole story. Nonetheless, it may be part of the story. I suggest that something like Hempel's account gives us much of the idea of what an explanation is – the idea of a *prima facie* explanation – something that *might* be an explanation. But whether a *prima facie* explanation really is in fact an explanation will be something that cannot be decided by a template. The issue is *holistic*, that is to say we need to consider all the *prima facie* explanations together and then choose between them. It is rather like the fair distribution of a cake. We cannot decide how much a person should have on a case-by-case basis; rather, we need to consider the number and strength of the competing claims before deciding how to divide the cake.

Before going on to show how we can explain the notion of explanation without a template or model, I want to raise yet another set of problems with Hempel's model. Looking at these will again point us in the right direction. The deductive–nomological model says that the law and conditions together *entail* the explanandum. The explanandum can be *deduced* from the explanans (hence the name of the model). This is a logical relation and that fact makes for problems. For one can add any amount of detail to the conditions and still preserve the entailment. This means that all sorts of irrelevant information can be added to the explanation.

The following satisfies the deductive–nomological model:

Law	Anyone who swallows 100 g curare dies immediately
Conditions	Bill ate a slice of cake on his birthday; the slice of cake contained 100 g curare
entail	———————————
Explanandum	Bill died immediately on eating the slice of cake

Here one of the conditions is that the cake was eaten on Bill's birthday. But that has nothing to do with explaining why he died (except indirectly in that it may explain why he ate the cake), any more than the fact that his birthday fell on a Tuesday or that Mr Nixon was President at the time. Any of these things could have been mentioned in describing the conditions pertaining at the time of Bill's death and could have been included in an argument like the one above that would satisfy the deductive-nomological model. Indeed that Bill ate cake is also irrelevant – all we need to know is the fact that Bill ingested curare. As long as we have that fact and the law, then the entailment goes through. The point is that irrelevant information can be made part of an explanation. In principle, anything may be made part of the explanation of anything else.

Such criticisms are not a real threat to the deductive-nomological model. After all, the very fact that the irrelevant information can be left out provides a way of improving the model. We might think that something should be part of the explanation only if it is essential to the deduction of the explanandum from the explanans. Even this proposal is not perfect, but rather than pursue that line I shall suggest that we should give up looking for ways to improve the model that retain the idea that the relation between explanans and explanandum is basically a logical relation of entailment. That Hempel should see explanation as such is not surprising. Hempel tacitly endorses the minimalist view of laws. On the minimalist view a law is just a generalization over its instances. The relationship between generalization and instance is simply one of entailment. But we also examined the full-blooded view that the relationship is more than this. Taking laws as necessitations among universals allows a different view of instantiation. Instantiation of a law is more than entailment. If the law is that Fness necessitates Gness then instantiation of this law involves something being both F and G. The fact that the thing may also be H is irrelevant, since Hness is not a universal that is part of the law. Instead of looking at the purely logical property of entailment we need to see what a law really is – a relation of universals. Something is explained by the law when that relation is instantiated.

Returning to our puzzles about explanation, they can be rephrased in terms of universals and the relations between them. Take the simplest case. Fness necessitates Gness – that is our law. We want to explain why x is G. This would seem to be explained by the fact that it is F along with the law. However, Hness also necessitates Gness and x is also H, so this could also explain x's being G. (Taking arsenic necessitates death, but so does being hit by a bus.) Which is the real explanation? Or another case, x is both

F and G, while the law states that Fness necessitates Gness, *and* vice versa. (A change in pressure necessitates a change in volume, and a change in volume necessitates a change in pressure.) So does being F explain being G, or does being G explain being F?

I propose that no model of explanation could answer this question completely. Hempel's model, reformulated in the guise of instantiation of universals, does nonetheless provide a model of *prima facie* explanation. This is a case where it is a law that Fs are Gs and something is both F and G, but it is not yet determined that it is G *because* it is F. So how do we know when something is not merely a *prima facie* explanation but a real one? Since there is no model of (real) explanation we cannot answer this question straightforwardly. The answer will depend on all the other *prima facie* explanations of the same fact that are available. Which is the real explanation will depend on which is the best of these *prima facie* explanations.

How, then, do we decide which is the best? To answer this we will look at a related problem, which the observant reader may have spotted. To have even a (complete) *prima facie* explanation we need to know what laws there are. On the criterial view, borrowing from the systematic account, the way we decide which laws there are depends on the best systematization of all the facts. The criterion for deciding that there is a law N(F,G) is that the facts are optimally systematized with a generalization that Fs are Gs. Systematizing a fact with such a generalization is the same thing as saying that the corresponding law explains it.

The criteria for the existence of certain laws and for the existence of certain explanations are closely bound up. One cannot decide what laws there are without also considering what explanations there are. The matter is settled holistically. That is, the criteria for $\{L_i\}$ and $\{E_j\}$ being the laws and explanations there are, are that $\{L_i\}$ and $\{E_j\}$ are those potential laws and potential explanations which all together make for the optimal Ramsey–Lewis systematization.

So the answer to the question of whether the law L explains the fact F must be this: there is no set of necessary and sufficient conditions for "L explains F". What we do have is criteria – we can say what sort of evidence is in favour of "L explains F". The criteria are the criteria for "it is a law that L" *plus* the following: F being captured/explained by L is part of what goes towards fixing the optimal system of which L is taken to be a part.

Inference to the Best Explanation

At this point we are going to jump ahead a bit. This section is about inference, and so strictly speaking belongs in Part II of this book. However, one of the key ideas in understanding scientific inference, *Inference to the Best Explanation,* is best introduced close to our discussion of explanation.

As we have seen, full explanations involve natural laws. So it would seem that in order to know that E is the full explanation of F one would need first to know the laws contained in E. Even to know that cause C is the incomplete explanation of D is to know that there is some law, which together with C and other relevant circumstances, makes for a complete explanation of D. So knowledge of an explanation requires some knowledge of what laws there are and so shares the same problems of induction from which knowledge of laws suffers.

While it is true that knowledge of explanations entails some knowledge of laws, it follows from this that explanation inherits Humean problems from knowledge of laws only if knowledge of laws is required before one can have knowledge of explanations. But perhaps this relationship may, on occasion at least, be the other way around. If it were, there would be occasions when we first know that E is an explanation and therefore know that L contained in E is a law, or that the event C contained in E as a causal condition actually occurred.

The idea here is that there are certain phenomena P in which we are interested. We think of possible explanations of P. Putative explanation E is the best available explanation of P, so we infer that E is the actual explanation of P. The putative explanation E may make reference to particular facts C and laws L, and so in inferring that E does explain P we are inferring that C and L are indeed actual facts and laws. This is known as *Inference to the Best Explanation.*

Inference to the best explanation is used extensively in science. This is most clear where the knowledge can be represented as an answer to a *why* question. Two examples: Why did the dinosaurs become extinct? Why is the universe expanding? Considering first the fate of the dinosaurs, was it that they were in the long run ill-adapted in comparison with their competitors? Or were they destroyed as a result of a catastrophe, such as meteorite impact? The latter hypothesis now has widespread, though by no means unanimous, support. But this is not because there is an observed correlation between meteorite impact and extinction – nothing of the sort has been observed directly. The lack of a Humean regularity does not prevent us from making inductive inferences. The hypothesis is that a large meteorite impact would have had such severe meteorological effects, mainly as a result of a cloud of dust in the atmosphere, that the dinosaurs were killed off fairly rapidly. The reasons for believing this hypothesis are: (a) the argument that if such an event had occurred then it would have the effect claimed – the extinction of the dinosaurs; and (b) two additional sorts of evidence, showing first that meteorite impacts do occur, and secondly that such an impact occurred at the appropriate time in history. The latter includes discoveries such as unusually high iridium deposits, which can be geologically correlated with times of known extinctions. Iridium is a common element in meteorites but not in the Earth's crust. This is good evidence for the hypothesis, since it adds to the facts the hypothesis is able to explain. Meteorite impact explains not only dinosaur extinction but also iridium anomalies. Furthermore, it explains why they are correlated. The explanation seems a good one – it explains diverse phenomena, and explains them well.

The other sort of evidence is evidence to the effect that events such as meteorite impact do indeed occur, i.e. that our explanation is not only good but also reasonably likely. This evidence includes counts and measurements of meteorite craters on Earth and the discovery of recent craters on the Moon. This evidence is helpful, but by no means crucial. It is important to note that thinking that an explanation is likely is not the same thing as thinking it is a good explainer of the explanandum. I may explain the extinction of the dinosaurs by hypothesizing a hurricane. That a hurricane occurred before the demise of the dinosaurs is highly likely – but it is still a poor explanation because it is difficult to see how it could have killed all the dinosaurs. Conversely, we can explain unlikely events and may have recourse to unlikely explanations to do so. The explanation can still be a good one. And if it is a very good one then that is in itself a good reason for thinking that it did indeed happen, even if it is a rare occurrence. The likelihood of an event is not on its own a good reason for thinking it explains another, although that it is a good explanation is itself a reason for thinking it more likely than one would otherwise have thought.

This is the case with the Big Bang and the origin of the universe. We cannot estimate the likelihood of an event like the Big Bang in the way we might estimate the likelihood of a hurricane, i.e. by observing the frequency of that kind of thing. But that does not prevent us from hypothesizing about it versus the alternative hypothesis of a steady-state universe. Nor does its uniqueness preclude us finding evidence in favour of its occurrence – such as background heat radiation across space, the relative abundances of hydrogen, Olbers' paradox, and Hubble's law (which is itself inferred as the best explanation of the red shift of distant galaxies). Insofar as a Big Bang theory can successfully explain these things, we may be in a position to know what happened when the world was born.

So far we have looked at explanations that are particular events and causes. But the best explanation may also be a law. It is perhaps tempting for the minimalist to see Newton's laws of motion and gravity as the tracing of patterns in the motions of things. I do not think this is right. It may be right for Kepler's laws, each of which does, separately, record regularities in the behaviour of the planets. This is a far less convincing picture with regard to Newton, as none of his laws individually records anything that can be observed. It is only from combinations of Newton's laws that we can derive the measurable motions of bodies. A more convincing rationale for Newtonian mechanics is that the laws together explain the observable motions of bodies, including Kepler's laws and otherwise unexplained facts, like the fact that unsupported terrestrial bodies fall to the surface of the Earth while the Moon does not.

I have described the general structure of inference to the best explanation, contrasting it with Humean induction. It is clearly at least an important component of our inductive practices and covers not only the inference of causes but also of laws. But important questions remain. What makes one explanation better than another? Why should we infer better explanations rather than poorer ones?

What then are the desiderata that make for a good explanation? Remember that most of the explanations we actually work with are incomplete, elliptical explanations, in which case we are taking something of a gamble on those bits that have been left out. In each case we have to ask: Is this a big gamble, or is it plausible that the gaps could, in principle, be filled? If we see a cup break on hitting the floor, that it hit the floor is a good explanation of its breaking since there is little doubt that the relevant extra factors (the fragility of the china, the force of hitting the floor, and so on) could, in principle, be incorporated into a full explanation. On the other hand, saying that a hurricane was caused by sunspot activity is, as it stands, a weak explanation, because without any further details it is not clear how one would fill this out to achieve a complete explanation. We do not know what would be needed to fill in the picture, let alone whether it is plausible that these things are true. This is why finding mechanisms is important in science. A causal hypothesis may well be entertained if there is some correlation between the two kinds of event. But the hypothesis is unlikely to be accepted as reliable unless some mechanism is supplied (or it is shown that the causal connection is an instance of a causal law). Wegener's theory of continental drift was not accepted for several decades because there was no clear idea of how it might work (Wegener's own ideas on this were wrong). It was not until the theory of plate tectonics was developed, providing a mechanism for continental drift, that Wegener's theory became widely accepted. For this reason, among others, parapsychological explanations are not regarded as credible. While there is some evidence of appropriate correlations, the lack of any plausible mechanism for telekinesis and telepathy keeps parapsychology at the level of speculation.

A second desideratum of a proposed explanation is its power. Good explanations explain several distinct facts, the more disparate the better. And they also explain in detail. For this reason quantitative explanations tend to be better than purely qualitative explanations. One of the strengths of Newtonian mechanics compared to its predecessors was its ability to explain with precision the orbits of the planets, not just their general form. The power of an explanation lies not only in its precision but also in its generality. If an explanation explains a lot of facts, then that counts in favour of that explanation. We have seen that the reason why the meteorite impact explanation of dinosaur extinction is a good one is that it also explains the quite distinct phenomenon of iridium anomalies. Similarly, Newtonian mechanics is a good explanation of planetary motions as it also explains terrestrial motion, whereas beforehand the two were thought to be unconnected.

A third good-making feature of explanations is their simplicity. An explanation of an event may employ many laws and a whole host of particular conditions, while a competing explanation may require few laws and conditions. Again Newtonian mechanics triumphed over its predecessors as an explanation of celestial motions because it was able to explain them with four laws, whereas both the Ptolemaic and Copernican systems employed vast arrays of epicycles.

A further positive quality of explanations is their ability to integrate or combine with other explanations. This idea is related to the power of an explanation. I mentioned that good explanations explain a lot. One way in which an explanation does this is by playing a role as a part of some other explanation.

The gas laws can be explained by the molecular theory of gases plus the kinetic theory of heat. The full explanation also employs Newton's laws of motion. This explanation is a good one because it integrates distinct laws, but it also supports other explanations that those hypotheses are used to make (e.g. the explanation of the result of Joule's experiments, which showed how kinetic energy may be transformed into heat). The idea here is that explanations can give each other mutual support by sharing common laws and conditions.

That these are the good-making features of explanations is attested to by the practice of scientists. It was Einstein's view that unity and simplicity are essential features of true theories.[19] Inference to the Best Explanation says that, like Einstein, we should be more ready to infer good explanations than poor ones, and a really good explanation that is clearly better than its competitors is likely to be the correct one. But why is this so? Why should we prefer better explanations? Why are they more likely to be true? Can we be sure that the world is a place where simple, strong explanations are true, and complex weak ones are false? And, if we cannot be sure that this is so, is it not irrational to put our faith in the one sort of explanation rather than the other?

You might notice that Hume's problem arises here again in another form. It cannot be a matter of logic that good explanations are more likely to be true. We might instead argue that our experience is that good explanations tend to succeed better than poor ones, and that the best explanation of this experience is that the world is so constructed that good explanations are more often true than are poor ones. With Hume's problem in mind we notice a circularity in that argument – we have used the technique of inference to the best explanation to justify inferring that our best explanations are more likely to be true.

A full resolution of this problem must wait. We will return to it in Chapter 4 and in Part II where I deal with questions of knowledge and rationality in science, but a few preliminary remarks will not go amiss. First, the question of whether it is reasonable to believe a hypothesis is not the same question as whether a hypothesis is likely to be true. The two questions must be somehow connected, as part of rationality at least is a matter of trying to get our beliefs to match what is in fact likely to be true. On the other hand, they are not identical since it can be rational to believe what turns out to be unlikely or false – for instance when we have bad data. For the moment I shall make some remarks about why it is rational to prefer good explanations. The reader may already have noticed the reason why. The good-making features of explanations correspond to the criteria for lawhood, as discussed in Chapter 1. Let us recall what being a criterion means. Say that C_1, ..., C_n are criteria for the existence of something, L. Then if C_1, ..., C_n are satisfied and countervailing circumstances do not arise, then we are rationally entitled to infer the existence of L. What did this mean for laws? The simplest criterion for the existence of a law is the existence of instances of the law. Our discussion of laws showed that this could not be all there is to it. First, because there can be laws with no instances, and secondly because what appear to be instances of this particular law may in fact be instances of some other law. Our understanding of laws has to be systematic. Whether a regularity is a law depends on how it relates to other laws. Correspondingly, our evidence needs to be considered systematically too.

The criteria for L being a law should be that the strongest, simplest systematization of all the facts there are should include L as an axiom or consequence. Evidence in favour of L are facts the knowledge of which makes a difference as to whether we should include L in our optimal systematization of what we know. This is why instances tend to be good evidence, since they make the most difference to whether we want to include L in our system.

Recall, further, that for *i* to be an instance of L, is for L to explain *i*. For the systems of laws to capture a set of facts is for those facts to be the instances of these laws, i.e. to be explained by them. So, to say that a criterion for R being a law is that R should be part of the optimal system, which is to say that R should appear as a law in the optimal system of explanations. This we saw earlier in this chapter. Optimality means strength, simplicity, and integration. So our optimal system is the simplest, strongest, integrated system of explanations. This is why we should prefer our putative explanations to be powerful, simple, and integrated with other explanations, as these are the criteria for something being a member of the set of actual explanations.

The hypothetico-deductive model of confirmation

Inference to the Best Explanation suggests that there is a link between explanation and confirmation (confirmation being the relationship between evidence and the theory for which it is evidence). How one will understand the notion of confirmation will depend on how one understands explanation. For a long time the hypothetico-deductive model of confirmation had currency as the counterpart to the deductive–nomological model of explanation. According to the hypothetico-deductive model evidence *e* confirms hypothesis *h* if *e* is deducible from *h*. Let our hypothesis be that all bats are blind. One may deduce from it that if *x* is a bat, then *x* is blind. So a bat that is found to be blind confirms this hypothesis. At a more sophisticated level, Eddington calculated the consequences of Einstein's theory of relativity for the observed position of certain stars during a solar eclipse. He found that the observed positions were as predicted (but not as predicted by calculations from Newton's laws) and this confirmed Einstein's hypothesis.[20]

The hypothetico-deductive model has intuitive appeal. In essence it says that if the predictions of a theory match what is observed then those observations confirm the theory. As the Eddington example shows, that is precisely what scientists seek to do. Nonetheless, the model suffers from severe defects. The best known of these is the *paradox of confirmation*. We saw that the hypothesis that all bats are blind entails that if x is a bat x is blind, and so is confirmed by Horace who is a fruit bat and who is blind. The same hypothesis also entails that if y is not blind then y is not a bat. Henrietta is not blind, nor is she a bat (she is a sharp-eyed red-setter). According to the hypothetico-deductive model, both Horace and Henrietta confirm the blind-bat hypothesis. In Horace's case, it is deducible from the hypothesis that if he is a bat he is blind, which is true, and in Henrietta's case it is deducible from the hypothesis that if she is sighted (not blind), she is not a bat, which is also true. Clearly, although Horace might confirm the hypothesis, Henrietta is completely irrelevant to it. (The paradox of confirmation is also called the raven paradox and Hempel's paradox, as Hempel first framed it in terms of the hypothesis that all ravens are black, which would be confirmed by a white shoe or, for that matter, by a red herring.)

Some philosophers, including Hempel, have attempted to save the hypothetico-deductive model of confirmation from this paradox by accepting the conclusion that Henrietta does confirm the hypothesis. The claim is that the sharp-eyed setter confirms the hypothesis but only to a very small degree. This can be supported by some probabilistic reasoning. Nonetheless, this cannot be right, or at least not the whole story. First of all consider that the hypothesis that all mammals are sighted is inconsistent with the hypothesis that all bats are blind. It is therefore plausible that evidence that confirms the hypothesis that mammals are sighted, thus disconfirms, to some extent, the hypothesis that bats are blind.[21] According to the hypothetico-deductive model, Henrietta does provide some degree of confirmation of the hypothesis that all mammals are sighted – she is a sighted mammal. And so she must therefore indirectly *disconfirm* the hypothesis that all bats are blind.

Consider Herbert, who belongs to a species closely related to bats. Herbert is very like a bat in many ways, genetically, environmentally, and so on. One difference is that he is sighted. According to the defenders of the hypothetico-deductive model he confirms the hypothesis that all bats are blind, since he is a non-blind non-bat like Henrietta. A geneticist might reason differently. They might think that because Herbert is genetically very close to being a bat, not to mention sharing the same habitat, it is likely that there are some bats, those which are genetically closest to Herbert, who share his power of vision. Thus in the context of further information, a non-blind non-bat might disconfirm the hypothesis that all bats are blind. The conclusion ought to be that whether evidence *e* confirms a hypothesis *h* depends not merely on the logical relations between *e* and *h* but also on all the other information that may be available. Confirmation is holistic. This parallels the impossibility of saying whether an explanation explains its explanandum, merely on the basis of their logical relation, independently of other possible explanations and relevant information. It also parallels the systematic nature of laws and the consequent impossibility of saying whether a regularity is a law independently of what other regularities there are.

The link between the failings of the hypothetico-deductive model of confirmation and the failings of the deductive–nomological model of explanation can be illustrated as follows:

Law	All bats are blind
Conditions	Henrietta is not blind
entail	______________________
Explanandum	Henrietta is not a bat

This satisfies the requirements of the deductive–nomological model.[22] Yet it is clear that the fact that Henrietta is not a bat is not explained by her not being blind or by the law that bats are blind. On the other hand, the view of explanation I have been promoting in this chapter does not fall into this error. If there is a law that all bats are blind, then it is the property of being a bat that necessitates blindness. Bats that are blind instantiate this relation, but non-blind non-bats do not. So Henrietta the sighted setter does not instantiate the law, and so her being a setter cannot be explained by it. Correspondingly, the hypothesis that all bats are blind cannot be confirmed by this case. If it were a law that bats are blind, that would not explain Henrietta's being a setter. So Inference to the Best Explanation cannot get to work here and no confirmation is provided for the hypothesis.

In summary, the deductive–nomological model of explanation and the hypothetico-deductive model of confirmation fail for the same reason – there just is no possibility of a simple model of explanation/confirmation in terms of a logical relation between law/hypothesis and explanandum/evidence. Instead our understanding of both confirmation and explanation must be holistic. The best evidence for a law L being the *explanation* of fact F is that F instantiating L is part of what goes to make up the optimal RL systematization, which is at the same time also the best reason to think that F *confirms* the hypothesis that L is a law.

Further reading

Carl Hempel's views on explanation are clearly articulated, along with his views on other topics covered by this book, in the paper "Explanation in science and history" and the short book *Philosophy of natural science*, as well as in greater detail in *Aspects of scientific explanation*. A contemporary discussion of explanation, including criticisms of Hempel, is David-Hillel Ruben's *Explaining explanation*. Inference to the Best Explanation is carefully analyzed in Peter Lipton's *Inference to the best explanation,* in which he also discusses explanation and the problem of induction. (He takes a view of explanation different from mine.) A useful collection is Ruben's *Explanation*.

Chapter 3

Natural kinds

Kinds and classifications

Philosophers and scientists have long bothered themselves with a pair of related questions: What is the constitution of things? What kinds of things are there? The questions are related in this way. If there are some basic kinds of things, then the differences between less basic kinds of things can be explained in terms of the natures of the basic kinds. So, for instance, the ancients followed Empedocles in acknowledging four elements (earth, air, fire, and water) and attributed the existence and properties of different kinds of things to differing proportions of these elements. It was this picture that encouraged alchemists to think that they could transform base metals such as lead, mercury, and tin into more valuable ones such as silver and gold. For all its faults, their outlook has been vindicated in its most general features by modern physics and chemistry. Being chemical elements, these metals cannot be transformed into one another by chemical means. But they do have common constituents in terms of subatomic particles (electrons, neutrons, and protons) and the transformation of lead into gold has in fact been achieved by physical means.

Despite the vainglorious desire for transmutation, the ancients and medievals nonetheless also believed that nature itself has fixed the kinds of things there are. Even if one could turn lead into gold, it is still the case that lead is lead and gold is gold and that there are essential differences between the two. The modern chemist would largely share this view. Nuclear decay notwithstanding, the elements represent discrete and more or less stable kinds of things. So too do most of their compounds.

Illustrating the notions of *element* and *natural kind* is one thing, giving a thoroughgoing explanation of them is another. For very soon we stumble across difficult cases. One of these involves the idea of sibling species. Creatures belong to sibling species if they are morphologically (i.e. anatomically)[23] indistinguishable from one another in the way one would expect of members of the same species. But in fact they do not all belong to one species, as is shown by the inability of subpopulations to interbreed. So we have distinct species, as defined by interbreeding, that are morphologically extremely similar and which can only be told apart with extreme difficulty (e.g. in the case of certain birds and frogs, by careful analysis of the calls they make). One case of this concerns the plant genus *Gilia*. *Gilia inconspicua* contains four sibling species that are morphologically identical and can be differentiated only by microscopic examination of their chromosomes. At the same time other plants of the same genus all belong to one species, *Gilia tenuiflora*, but are readily seen to fall into four morphologically different types.[24] It is quite unclear how we ought to divide up these plants into kinds. If we use the genetic/interbreeding species concept, then we will classify four obviously distinct sorts of plant as belonging to one kind. On the other hand, if we employ an anatomical principle to distinguish between different kinds, one of those kinds will comprise four subgroups divided from one another by intersterility.

Perhaps a more homely example might allow us to get an easier grip on the difference between natural and unnatural ways of classifying things. When one visits a greengrocer, in the section devoted to fruit one will find, among other things, apples, strawberries, blackcurrants, rhubarb, and plums, while the vegetable display will present one with potatoes, cabbages, carrots, tomatoes, peppers, and peas. If one were to ask a botanist to classify these items we will find rhubarb removed from the list of fruit and tomatoes and peppers added (these being the seed-bearing parts of their respective plants).[25] We have two ways of classifying the same things, one biological and the other culinary. If we look at the defining features: "part of a plant either eaten uncooked, for its sweet taste, or cooked with sugar in the preparation of puddings and deserts or other sweet dishes" versus "the seed-bearing part of a plant" the former may appear rather too related to (modern, western) human interests, while the latter refers to its biological features. It would seem clear that the latter is more natural.

Further reflection, however, muddies this apparent clarity. The "biological" definition is more natural only if the terms it employs are themselves natural. That is, if "seed" happens to be an unnatural, anthropocentric concept, then "seed-bearing part" is going to be no better. So we had better be sure that "seed" is a natural concept. To assure ourselves of that we need to look at its definition and check that its definition and component concepts are natural too. But what would that achieve? We have merely shifted the problem from the concept "fruit" to the concept "seed" and thence to the concepts used in its definition.

Following this line of reasoning one might conclude that there really is no absolute sense in which there is a natural classification of things into kinds. For sure there are classifications that we use for science and others that we use for different purposes. Yet, could not other classifications be used for science? We are taught that whales are not fish but mammals. But might we not have used an environmental classification and kept whales with sharks and other large sea dwellers? The classification of chemical compounds is equally open to choice and variation. We may choose to classify organic compounds according to the basic compound from which they are derived, or according to their active groups. So we may put C_3H_7COOH (butanoic acid) with $C_3H_7CH_2OH$ (butanol), while placing C_6H_5COOH (benzoic acid) and $C_6H_5CH_2OH$ (benzyl alcohol) together since the first pair are both derivatives of butane and the second pair are both derivatives of methylbenzene. Alternatively, we can regard C_2H_5COOH and C_6H_5COOH as similar since they are both carboxylic acids (active group COOH), while $C_2H_5CH_2OH$ and $C_6H_5CH_2OH$ may be categorized together in virtue of their being alcohols (active group CH_2OH).

It is now perhaps not so clear whether there exist natural kinds, and if so what they are. In the end is our classification merely *our* classification? Or might it match natural divisions and distinctions among things? To employ Plato's much-quoted (because striking and apt) metaphor, does, or might, science carve nature at the joints? Or indeed, might nature have no (discernible) joints at which to be carved?

The descriptive view of kinds

The issue was initially couched in terms of finding a natural definition for a certain kind of thing, or a natural principle of classification. To take the debate further we will need to distinguish three aspects of a kind concept. First, there is its *sense*. This is a description of the properties that a thing must have in order to be a member of this kind and such that if a thing does have these properties then it is a member. The sense is something which a competent user of the expression must know. The sense is what someone grasps who understands the term in question. So for "tiger" the sense might be: carnivorous, four-legged, black and yellow striped, feline, large mammal. (The sense is sometimes called the *intension*.) Secondly, there is the *extension*. This is the set of things which are members of the kind, that is all the things that possess the properties making up the sense. In this case, the extension of "tiger" consists of all tigers. Finally, the *reference* is the abstract entity the natural kind (or universal) *tiger*.

This approach to natural kinds I shall call the *descriptive* view. The obvious problem with the descriptive view is that, as it stands, natural kinds have too many possible exceptions. A tiger that has lost one leg is still a tiger, and so is an albino. To meet this objection it has been proposed that the condition for membership of the kind be relaxed. It is not that something should have all the properties on the list to be of the kind but only sufficiently many. So a three-legged tiger is a tiger if it has almost all the other properties associated with being a tiger (black and yellow stripes, carnivorous, etc.). If something has too few of these, then it is not a tiger. (This is the *cluster* view.)

Along with this view it is often held that our natural kind concepts change over time in step with advances in scientific knowledge. So the sense of "water" would once have been that water is an odourless, colourless liquid that dissolves salt but not gold, rusts iron, and so forth. Over the years, from Lavoisier on, it was discovered that what was called "water" under this definition also has the chemical structure H_2O. This discovery, once established, provided a new definition. Nothing would be lost, since everything which is an odourless, colourless liquid, etc. (i.e. everything that satisfies the old definition), also has the structure H_2O. So the extension of the old concept and of the new are identical. At the same time the new definition had various advantages – it is simpler and more definite than the old definition and it is more useful to those researching into chemical structure. The sense of "gold" has similarly changed, from that of a shiny, yellowish, dense metal, largely inert but soluble in aqua regia to that of the element with the atomic number 79. This explains how what was a discovery can now be a definition. Such changes in our concepts are in no way forced upon us but are a matter of our convenience and interests. The idea is that for such reasons a word may change its sense/intension over time so long as care is taken to preserve the extension.

Intensions of this sort describe what is called the *nominal essence* of a kind. The nominal essence of a kind K consists of those features a thing must have to deserve the name "a K" by virtue of the meaning of that name. So the nominal essence of a knife is that it is a hand-held implement with a single cutting blade. If the descriptive view is right, then natural kinds have nominal essences. The nominal essence of water is given by the intension of "water", and so the nominal essence was once that it is a clear, colourless liquid, etc. It is important to bear in mind that because nominal essences are intensions they should typically be known by competent users of the expression. According to the nominal essence view, if one knows what "water" means, one should know that water is a clear colourless liquid.

An essentialist view of kinds

The descriptive view just outlined has been subject to penetrating criticism. If the criticism is right, then natural kinds do not have nominal essences. The first argument shows that having a nominal essence is not necessary for a natural kind, the second argument says that it is not sufficient.

The first argument comes from Saul Kripke. The nominal essence we suggested for a tiger was: a carnivorous, four-legged, black and yellow striped, feline, large mammal. So a creature must have all these properties to be a tiger, or most of them on a cluster view of nominal essence. Kripke argues against both of these, by suggesting that we can conceive of circumstances such that tigers do not have any of the properties we take to comprise the nominal essence. He imagines that our knowledge of tigers derives solely from the reports of explorers who have seen tigers in the jungles of India and surrounding lands – they have reported seeing these elusive animals (which they have been unable to capture and examine in detail). Calling them "tigers", they say that they are striped, black and yellow, are large, fierce and carnivorous, four-legged, and so on. On the hypothesis that tigers have nominal essence, they must consist of these properties since these are the only ones anyone knows about. But, argues Kripke, tigers may not have even these properties.[26] It might have been that it is a trick of the light in the jungle that the animals seen by the explorers appeared to be striped black and yellow; in fact these animals are blue or some other colour. A similar trick of the light might have caused the observers to miscount the legs – in fact the animals they were looking at had five legs (one could think of alternative stories; perhaps normally tigers have five legs, but those which had accidentally lost a leg were slower and were thus the ones seen by the explorers). Perhaps by nature tigers are very timid and are herbivores, only that in one or two cases the tigers had an allergic reaction to the explorers' pipe smoke and went crazy. Indeed, one could imagine that these tigers were not even animals – had one been captured (which was not achieved) and dissected it would have been seen that they are robots and not animals at all (and hence no natural kind).

The point of these tales is that, even if far-fetched, it is not inconsistent to assert that the tigers seen by the explorers had none of the properties they appeared to the explorers to have. So what of the nominal essence? It cannot be "black and yellow striped, four-legged carnivorous animal", since tigers do not have this essence. Indeed it cannot be anything, since there are no interesting properties that are both possessed by the tigers and believed by the users of the name "tiger" to be possessed by tigers. So tigers have no nominal essence. But we do not want to say that there is anything wrong with talk of tigers. What they say may be wrong, but people who talk of these marvellous creatures seen by the explorers are nonetheless talking about those blue coated, five-legged herbivores – tigers. And subsequent explorers may find out that what was first believed about tigers was all wrong. An actual case might be found in whales, of whom people would for a long time have given the purported nominal essence "large water-spouting fish". The fact that whales are not fish does not mean that such people were not talking about whales at all. Rather, they were talking about whales – only whales do not fit the suggested nominal essence, which just shows that it is not an essence at all.

Kripke's example shows that a nominal essence is not necessary for the existence of a natural kind. Examples from Hilary Putnam show that a nominal essence is not sufficient either. His strategy is to imagine two worlds in which the state of knowledge and belief is identical, but in which the natural kinds being talked about are different. In which case the nominal essences are insufficient to pick out natural kinds.

Putnam's two worlds are the Earth as it is and Twin Earth, which is like Earth in most regards except that where Earth has H_2O, Twin Earth has a different compound, denoted by, for the sake of argument, the formula XYZ. H_2O and XYZ function identically in all superficially observable respects. XYZ is a colourless, tasteless liquid, which boils at 100°C and freezes at 0°C. XYZ supports life on Twin Earth, is what constitutes precipitation, fills the rivers and lakes, and comes out of the taps to fill the baths. Today we can distinguish between Earth water and Twin Earth water because we have the chemical knowledge that allows us to see that the underlying chemical constitution of the two liquids is different. So what we on Earth call "water" is a different kind from that which our counterparts on Twin Earth call "water". However, in 1750, the relevant chemical knowledge then lacking, inhabitants of the two planets would not have been able to tell the two substances apart. Putnam argues that there would still have been two substances. The word "water" as used by English speakers on Earth in 1750 would have referred to what we now know to be H_2O, while their Twin Earth counterparts would have been referring to XYZ. Because of their limited chemical knowledge, neither the Earth people nor the Twin Earth people are able to provide satisfactory general descriptions that would distinguish what they call "water" from what is called "water" on the other planet. Both populations would assent to propositions such as "water is a colourless, tasteless liquid, which boils at 100°C and freezes at 0°C", etc. There is no description they would disagree on. Hence the nominal essence of water must be the same for both sets of people.

The extension of "water", however, differs between the two groups, one covering things that are H_2O and the other covering that which is XYZ. It follows, therefore, that the nominal essence is not sufficient to fix the extension of the term "water".

Putnam instead has another view of what fixes extension. He says that associated with a kind term are archetypes, i.e. examples of the kind. These serve to fix the extension of the kind term via something called a *same kind relation*. Two things have this relation if they are of the same kind; that is to say if they are identical in relevant scientific respects – typically that they behave in nomically identical ways, e.g. if substances react in the same way in chemical reactions and have the same general physical properties (density, melting and boiling points, and so on). In turn this is founded on their having the same microstructure – this being the chemical composition for a substance, and something more complicated such as a detailed molecular structure for the genes underlying the sameness of species.

As the same kind relation involves underlying structure and properties, two things may have this relation even if they are not known to have this relation. And so, in 1750, only things that are H_2O would bear the same kind relation to an archetype that is H_2O, and buckets of XYZ would not have this relation to that archetype, even though no one at the time could have known this. And so someone on Earth who points to a sample of rainwater and says "water is whatever is like (has the same kind relation to) this" will thereby be picking out only H_2O and not XYZ. The same will be happening on Twin Earth. We can now see what has happened with Kripke's tigers and with whales. The tigers seen by the explorers were the archetype for their word *tiger*. What makes something a tiger is its genetic structure. The description they gave of tigers and which others used could be completely wrong, since animals sharing that genetic structure might not have that description. The description "large water-spouting fish" should not be regarded as giving the nominal essence of *whale*. Rather it is an empirical and false description of whales. What makes a creature a whale is that it shares the genome of those actual creatures encountered by sea farers and called "whale" by them.

Kripke and Putnam have shown that natural kinds do not have nominal essences. More positively they have provided an account of what natural kinds really are. The basic idea employed in both their examples is that of a sample or an archetype, which defines a kind via a "same kind relation" – to be of the kind something must be just like the sample. What being the same kind amounts to is discovered by science – having the same molecular formula or having the same genetic structure. These will be properties that something must have to be of the kind, but unlike nominal essence will not be explicitly part of a verbal definition, will not be an intension. Such properties we may call the *real essence* of the kind. The real essence of water, therefore, is being H_2O, that of tigers is their genetic structure, and so on. This view is called *essentialism*, becase it maintains the existence of real essences, which it is the task of science to discover.

We need to distinguish between what we know or believe about the nature of a kind and what its nature really is. On the descriptive view the intension/nominal essences are known in advance of any scientific enquiry, and someone who has a proper grasp of the relevant concept cannot be wrong about these. By contrast one can be ignorant of the real essence of a kind – this is for science to discover. In either view the essence kind consists of those properties that a thing or piece of stuff must have to be of that kind. Those properties of the kind could not be otherwise, while some other properties might have been different.

Kripke gives another example that illustrates this. He says that we might have been mistaken in thinking that gold is yellow; it might really be blue (optical illusion again, or a strange property of air or light – think of photographic film, which changes colour when exposed). So being yellow could not be part of the nominal essence of gold. Similarly, we might have been mistaken in our investigations that tell us that gold has atomic number 79. So that too cannot be part of the nominal essence. But if it is true that gold does have atomic number 79, then anything, to be gold, must have atomic number 79, whatever the circumstances. Could something have a completely different appearance and still be gold? Under unusual circumstances it could, and it would still be gold by virtue of its atomic number.

In passing, an oddity with Kripke's essentialism may be noted. He says that gold necessarily has atomic number 79, but is not necessarily yellow. However, the atomic make-up of gold is what explains such features as its crystalline structure, its density, its colour, and so on. That is, it explains all the features normally regarded as gold's nominal essence. Thus, with the laws of nature as they are, gold must not only have the atomic number it has, but also the colour, density and malleability that it does. And so, contrary to what Kripke says, it is not the case that gold might have been blue. Of course, if the laws of nature were different, gold might have been blue, but then with different laws atomic and genetic structure might not have the same significance, indeed there might not even be such things. What Kripke is driving at is that when we are interested in the essence of a natural kind we are not thinking of just any properties this kind necessarily has, but those properties that are in some sense fundamental. So the atomic number explains the colour, but not vice versa, and the genes explain the physiology, but not vice versa. This is an important idea and we will return to it later.

Problems with Kripke–Putnam natural kinds

Kripke's tiger story is aided by the fact that explorers saw tigers only once. But this is unlikely for most natural kinds. By and large many samples will be seen – as in the case of whales and water.

What this means is that we will rarely be quite as wrong about kinds as his story suggests. So we will usually be able to use "superficial" descriptions of things to identify their kind reasonably accurately – so the superficial description given of tigers (striped, black and yellow, are large, fierce, carnivorous, and four-legged) will in fact typically pick out only tigers. This is what gives the description the semblance of a nominal essence. But Kripke and Putnam have shown that we might be mistaken; in particular, there might be more than one kind that shares the same superficial features. So what we thought of as being many samples of the same kind may in fact be nothing of the sort, but examples of different kinds. So the single covering name may have been used as if it were a natural-kind term, although it turns out not to be so; it transpires that it covers two kinds. A well-known example of this is jade. Jade comes in two apparently similar forms, jadeite and nephrite, which are very similar in superficial properties, such as colour and toughness, but which have quite different chemical compositions and crystalline structures.[27]

That what was thought to be a single substance turns out to be two or more different substances may be no surprise. But is does raise a question about the standard of similarity that is supposed to be employed. We may say "something is gold if and only if it has the same microstructure as this sample", but what is meant by "microstructure" is left unsaid. In the examples such as the case of H_2O and XYZ, those who make this stipulation are completely ignorant of any science of microstructure. For Putnam's case to work, people in the eighteenth century are supposed to have intended that their use of the word "water" was to cover only what was the same as the water they saw about them. What science subsequently shows "the same" to amount to is "having the same molecular structure". But the latter is a notion they could not have even entertained, let alone intended.

One way round this is to take "the same" to mean "absolutely the same in every respect". This is a catch-all phrase, which does the job of including sameness of molecular structure. But it is too catch-all, since most of the water anyone in the eighteenth century would have come across would have been impure. And so if anyone had a particular sample in mind then something would have to have precisely the same quantity of impurities in order to be what they called "water". We may therefore want to relax our notion of sameness a little, to mean "the same in all (scientifically) important respects" intending the impurities to be regarded as insignificant. This in turn generates problems: when is something water with an insignificant impurity and when is it a dilute solution of potassium cyanide? (Nor is the question trivial for gemology. Impurities make all the difference between a stone being a ruby and so very valuable and its being sapphire and so rather less so, though both are corundum (Al_2O_3), differing only in their colours, for which traces of iron, titanium, and chromium are reponsible.)

In the case of water, perhaps part of the answer is to reflect on the fact mentioned above, that no one ever really used a particular sample to fix their concept of water. Rather the archetype is not a single clearly isolated sample but a loose collection of samples indicated by something like Putnam's phrase "what we call 'water' round here". Clearly, among all such samples H_2O is going to be the dominant ingredient, and with luck the only common ingredient.

So perhaps we can come up with some reasonable catch-all concept of sameness, suitably modified not to be upset by impurities. What it does entail is that natural kinds will be perfectly homogeneous. It may be that two samples are, in this modified view, allowed to belong to the same kind, even if there is the odd impurity. But this is the case only if the dominant components are absolutely alike. There is no room for a slight difference, as a slight difference is still a difference. For example, we will not be allowed to regard chemical elements or compounds as natural kinds because of the existence of isotopes, isomers, and polymorphs, such as allotropes. Let us take the example of tin. Tin has 21 isotopes. Atoms are different isotopes of the same element if they differ in the number of neutrons in the nucleus (being the same elements they have the same number of protons). So some atoms of tin, along with their 50 protons, have 68 neutrons, while others have 69 neutrons, and yet others have more or less than either of these. It is possible, indeed likely, that two samples of tin will contain differing proportions of these isotopes. In an extreme sample one might be pure tin-118 and the other might be pure tin-119. The samples will therefore differ in their microstructure and so cannot be of the same kind according to the definition we have gleaned from Kripke and Putnam. So what we call the element tin is not a natural kind but a mixture of 21 different kinds, one for each isotope. Tin also affords another example of microstructural difference. This is the phenomenon of allotropy. Tin exists in two different forms: "white tin", which is metallic, and "grey tin", which is non-metallic. These differ in their micro(crystalline) structure and so also cannot be the same kind.

Isotopic and allotropic differences cannot be ignored in the way that impurities can. The extra neutrons in isotopes are part of the make-up of the stuff in question. Furthermore, they are not scientifically insignificant since their presence is important in explaining such phenomena as nuclear fission. The thought that every isotope should be a kind is not in itself objectionable. But it does restrict our notion of kind. It means that chemical elements are not kinds, but like tin are mixtures of them. Consequently, chemical compounds being made up of elements cannot be kinds. Water cannot be a kind if hydrogen and oxygen are not. Rather, it too is a mixture of kinds, one for each possible combination of different isotopes of hydrogen and oxygen. Ironically this means that Putnam's eighteenth century Earth and Twin Earth communities were not in fact picking out natural kinds at all. Indeed, it is very difficult to pick out natural kinds, except for the nuclear physicist, since everything of interest to chemists, biologists, and so on, will be made up of compounds of different isotopes.

Allotropy presents another problem. Different allotropes (unlike different isotopes) are so different as to differ to the naked eye. But in the case of tin they do not seem to be "fixed" in the way we expect of natural kinds. White tin is stable above 13°C, while grey tin is stable below that temperature. White tin can be turned into grey tin by cooling it (called "tin plague") and this process can be reversed by heating the grey tin to above 100°C. Say this occurs to a pure isotope of tin. Do we have here the change of one kind into another? Or is it just one kind appearing in two different guises? On the one hand, we might feel this is rather like a difference in state (liquid, solid, and gas), just different forms of the same stuff. On the other hand, one might associate it with a better known case of allotropy, that of graphite and diamond, where these seem not to be one kind. It simply is not clear what counts as a kind here.

A similar problem arises in biology. Kripke's tiger example implies that he thinks that tigers form a natural kind. But nothing could be more obvious than the fact that biological species do not have the sort of homogeneity that is required of a natural kind of Putnam's sort. What is the "microstructure" in the case of an organism? One could reasonably take either the physiological structure or the creature's genetic code. While the former is closer to the notion of a *structure*, Kripke, as we have seen intends the latter because it is causally more basic. The genetic make-up of an organism (its genotype) explains its physiological and biochemical make-up (the phenotype). Either way, the structures involved are massively complex, and every individual differs genetically and physiologically from every other. Reasoning parallel to that in the discussion of isotopes leads to the conclusion that every individual is its own natural kind. So Kripke's explorer, if he did pick out a kind, succeeded in doing so with respect only to the kind consisting precisely of the animal he saw and none other.

One moral that may be drawn from this discussion is that, although one might require absolute similarity among all members of a kind, this requirement might usefully be relaxed. While perfectly homogeneous kinds might form the most basic level of kinds, other levels might have members that differ in some respects but are similar in important properties. We might thereby have nested kinds. One kind could have two or more subkinds. So chlorine is a kind consisting of several subkinds, one for each isotope. The nesting of subkinds within broader kinds would be an appropriate approach to biological taxa. The hierarchy of groups of organisms into kingdom/phylum or division/class/order/family/genus/species/ subspecies or variety already displays this.[28] The idea of non-basic kinds can be extended to allow not only for the nesting of kinds one within another, but also for overlapping and criss-crossing kinds. We saw this in the different ways of classifying organic compounds.

Desirable as the idea of non-basic, non-homogeneous kinds is, it nonetheless raises again the question we started with. If we are to group things together that are not absolutely identical, then those things will share some properties in common and not others. How do we choose which properties are appropriate for this purpose? Which properties ensure a natural classification? Or is there no such thing beyond the most basic level?

Natural kinds and explanatory role

Before answering this question, I want to raise another challenge for the Kripke–Putnam view. It turned out that their account identifies only perfectly homogeneous classes as kinds, which in practice are going to be mainly single isotopes of elements plus subatomic particles. The mechanism proposed by Kripke and Putnam for naming a kind, which employs a sample functioning as an archetype, an anchor for the same-kind relation, seems inappropriate for these kinds. It is rare that anyone comes across a sample of a single isotope, and as for subatomic particles, do we really use samples of them as a basis for similarity? How does one obtain a sample neutrino and how does one test other particles against it for sameness? Granted, it is possible to produce and examine electrons – ever since the invention of the cathode ray tube (the forerunner of the television tube). But it is not the case that the notion of electron was introduced in this way; rather, the emissions produced by the heated cathode were called cathode rays. What happened was this. William Crookes hypothesized that these rays were composed of charged particles. "Electron" was the word used by Jonathon Stoney for the unit of charge carried by monovalent ions in electrolysis – resuscitating Faraday's notion of a unit of charge to explain electrochemical phenomena. J. J. Thomson demonstrated that Crookes was right about cathode rays and that the charge of the constituent particles is equal to the value of Stoney's "electron"; thereafter, "electron" became the name of the particle rather than the unit of charge.

Here we do not have any clear archetypes. Rather we have certain phenomena, electrolysis and cathode rays, in order to explain which discrete units of charge or charged particles were hypothesized. It is rather as if the phenomena are playing the role of archetype and the kind is that kind which explains those phenomena. Once such a kind has been discovered, its properties, such as its mass in the case of the electron, can be investigated.

This explains how we can introduce terms that are intended to be natural kind terms, but fail as such because there is no such kind. Phlogiston is a case. The eighteenth century chemist Joseph Priestley, among others, hypothesized the existence of a substance, phlogiston, to explain combustion. Phlogiston, they supposed, was a subtance contained in inflammable materials that enabled them to burn; it was given off in combustion, and air saturated with phlogiston would not support combustion; air that would allow combustion they called "dephlogisticated air", while Lavoisier claimed that it supported combustion not because of the lack of phlogiston but because of the presence of oxygen. There is no phlogiston and so there could be no samples of phlogiston. A straightforward Putnamian theory could not work for the meaning of "phlogiston". But the phenomenon in question, combustion, does exist, and so phlogiston could be hypothesized as that thing which explains combustion, as a substance contained in inflammable materials, and given off by combustion. As in the case of the electron we have a phenomenon and hypothesize a kind to explain it.

This example shows that we must say a little more than just that a substance is hypothesized to explain a phenomenon. After all, oxygen and electrons explain combustion, but we do not suppose that "phlogiston" refers to either of these (or both). When a kind is introduced as explaining a phenomenon, the kind of explanation expected may well be part of the kind concept. So phlogiston was expected to be a substance contained in inflammable materials.

The phenomenon in question may itself be highly theoretical, or may not have even been observed, and the degree to which the expected explanation is filled out may be considerable. So neutrinos were posited in order to explain deficits in masses after nuclear interactions. Positrons were predicted by Paul Dirac to exist as a consequence of his relativistic theory of the electron. There is a sense in which positrons were themselves not intended to explain any characteristic phenomenon, although such phenomena (e.g. the creation of an electron–positron pair) were later observed as expected. Rather, the theory as a whole is supposed to explain the behaviour of the electron. Here, the understanding of the notion of a positron is tied very closely to that theory and how it contributes to the explanation of quite general phenomena.

We are now getting to the heart of the matter. Natural kind concepts get their meaning by virtue of their explanatory role. What that explanatory role is may involve greater or lesser amounts of theory. Where we have a sample of the kind, the amount of theory may be very small indeed; in other cases the theory may be spelt out in detail. The connection between kinds and explanatory role allows us to make room for non-basic kinds, as we can forge explanations at the appropriate level in chemistry, microbiology, genetics, and so on, not just at the level of particle physics. Explanations also criss-cross. In explaining why benzyl alcohol reacts with phosphorus chloride to form benzyl chloride, we may point out that this is a reaction characteristic of alcohols, while if what needs an explanation is the same substance's volatility and odour, then reference to its being an aromatic compound (a benzene derivative) is called for.

We have also gone a long way to answering the original question, whether some classifications are natural. Some classifications are no good for explaining things. One might randomly collect diverse things and give the collection a name, but one would not expect it to explain anything to say that a certain object belonged to this collection. It is difficult to see what obvious natural laws or explanations employ the culinary notion of a fruit, and rather easier to see the explanatory significance of the biological notion. In the case of sibling species, biological opinion has historically turned away from morphological classification in favour of genetic classification. The reason, I suggest, is that the grouping of creatures into reproductively distinct units is far more explanatorily powerful than the alternative, especially when one is concerned with the long-term evolution of species. However, there may be a use for morphologically described groupings in different explanatory contexts, e.g. explaining the response of a population to short-term environmental changes. If so, it would be legitimate to allow such groupings to be kinds. The conclusion we ought to draw is that there are objective natural kinds, the existence of which rests on the possibilities of explanation, and so on the existence of laws of nature. Nonetheless, the concept of kind does not vindicate the view that for each entity there is only one natural kind to which it belongs. On the contrary, there may be many, the interrelations of which, especially for more complex entities and, above all, biological ones, may be very complicated. In some cases, one kind may be a subkind of another, while in other cases kinds may overlap without one being the subkind of another.

Laws and natural properties and quantities

Our discussion of natural kinds has consequences for our conception of laws. The picture of the world suggested by the minimalist about laws is one of objects and their properties and relations constituting the basic facts that there are, while laws, in a sense, piggyback upon these by being certain sorts of regularity among these properties and relations. This suggests that those properties and relations are independent of the laws – that in another possible world the same properties might be found in different regularities and so enter into different laws. So, for instance, it is an observed regularity that black objects warm up more quickly than white ones. We could conceive of things being the other way around. This is because we are able to identify the colour of objects and their temperature independently of having any inkling of their having a nomic relationship. This picture is a venerable part of empiricism, and is to be found in the philosophy of Hume and Berkeley as well as more recent philosophers. It is even present in the writing of anti-minimalist philosophers such as David Armstrong, who thinks that whether two properties are related by nomic necessitation is a purely contingent matter.

I think this picture is very misleading. As we have just seen, sometimes our only conception of a certain natural property or quantity is in terms of its explanatory role. As explanation requires the existence of a law (see Ch. 2), our conception of the property or quantity will entail some recognition of the existence of the law or laws in which it features. Take the quantity *electric charge* and the laws that like charges repel one another and different charges attract. It is not that people first had a concept of electric charge and then subsequently observed the regularity that like charges repel. On the contrary, it was noticed that under certain conditions objects can be made to attract and repel one another, and it was hypothesized (by Thales, Theophrastus, and William Gilbert) that this is a nomic rather than accidental feature of the world. Slowly the concept of electric charge was developed (by Du Fay, Priestley, and Coulomb) as it became clear that a certain quantity must exist which enters into the law thus identified. Note two features of this rather schematized and condensed history: (a) the discovery of the quantity came *after* the discovery of the existence of the laws it is part of; and (b) the concept of electric charge is dependent on these laws – what we mean by "charge" must be explained by reference to these laws. We could not have charge without also having these laws (or laws very similar to them). (I suspect that the same goes for concepts like *mass* and even *matter* – that what we mean by these terms depends on our identifying, even if in an unarticulated way, the existence of laws of inertia and gravity, in which case the idea of a world without any laws at all is incoherent. For there to be anything at all there must be some laws and some kinds.)

A problem for natural kinds

That natural kinds are to be understood in terms of their role in explanation is to be expected in view of the connection between explanation and laws. It will be relevant to mention the properties of a thing in an explanation if those properties are referred to in the laws that underwrite it. So to say that natural kinds play an explanatory role is to say that they appear in natural laws. Natural kinds correspond to the universals linked by nomic necessitation. The idea is that if F is a universal appearing in some natural law, then Fs form a natural kind. It may be odd to think of there being laws about tigers. If there are no laws about tigers, there are at least law-like truths that support explanations concerning tigers. A law-like truth may be something like "tigers are (usually) carnivorous" – such a truth will be supported by laws (e.g. of genetics); law-like truths may have exceptions because they also depend on non-nomic regularities with exceptions (e.g. environmental factors).

Having lighted on the idea that the universals related in natural law correspond to natural kinds and having seen its plausibility, we now need to reflect upon a problem. Let us remind ourselves then what we said about laws. The criterion for being a law is that it should follow from the simplest and strongest axiomatization of everything. We had a problem with simplicity. If we used different predicates we could turn a simple axiom into a complicated one, and vice versa. And so, in order to get a determinate notion of simplicity going, it looks as if we need to specify a set of predicates in advance.

As these are the predicates that appear in statements of natural law, they are going to be our natural-kind predicates. However, what I have just said about simplicity suggests that the choice of these predicates is arbitrary. At the very least, it means that the characterization of natural kinds as the kinds corresponding to the universals in natural law, although true, will be trivially true and useless, because it turns out that what the natural laws are depends on which the natural kinds are. There is a circularity here between the notions of natural law and natural kind.

Circularity in definitions is not necessarily a very bad thing. It may be that we have a set of interrelated notions none of which is basic. In understanding any of these notions fully we have to grasp the whole lot and their relations together. This structure, perhaps with other features governing the use of these concepts, such as canonical examples of their use or samples of their instances, may be sufficient to give a reasonably determinate content and application to them.

The fear is that we do not have this determinacy here. Let us accept, for sake of argument, that it is a law that sapphires are blue and a law that rubies are red. Let us make the following definitions:

x is a sapphuby if and only if:

either x is observed before 2001 and is a sapphire,
or x is not observed before 2001 and is a ruby.

x is a rubire if and only if:

either x is observed before 2001 and is a ruby,
or x is not observed before 2001 and is a sapphire.

x is bled if and only if:

either x is observed before 2001 and is blue,
or x is not observed before 2001 and is red.

x is rue if and only if:

either x is observed before 2001 and is red,
or x is not observed before 2001 and is blue.

From the facts that sapphires are blue and that rubies are red, it follows that rubires are rue and sapphubies are bled. Furthermore, the reverse also holds; it follows from the facts that rubires are rue and sapphubies are bled that sapphires are blue and rubies are red. So the two pairs of generalizations are equivalent.

The first question is: Are the two new facts laws? From the perspective of one who has just constructed the two facts out of the two laws about sapphires and rubies, these are law-like facts, but not laws. The reason is that a disjunction of properties is not always itself a property; at least it is not a natural kind property.

However, correct as this claim may be, it is not entirely satisfactory as an answer to our question. We have said that it is not a law because being a sapphuby is not a natural property. But why not say that being a sapphuby is a natural property, *and* that it is a law that sapphubies are bled? One reason why not might be that being a sapphuby does not explain something being bled. There is a correlation, but this is by virtue of two quite independent explanations, of something being blue if it is a sapphire and of something being red if it is a ruby.

Nonetheless, a philosopher (or scientist) determined to hang onto the natural kind of sapphubies might say that this perspective has been adopted only because we started out with a vocabulary of sapphires and rubies, and blue and red. As we were going to frame our laws in this vocabulary, we had in effect decided that these were going to be our natural kinds, and hence had decided in advance to exclude sapphubies from being a natural kind. If, however, we had started out with a vocabulary of sapphuby and bled, we would have come to a different conclusion. From a neutral perspective the two systems appear equally strong and equally simple.

It looks as if the problem is not an epistemological one (that we cannot tell which the laws and natural kinds are). It appears to be deeper than that, concerning the very concepts of law and natural kind. The explications we have given of the two are so circular that, taken together, the pair of concepts do not have sufficient content to make it determinate what we mean by "law" and "natural kind". Expanding the sorts of argument canvassed above makes it look as if there could be an infinite range of systems and properties all competing to be regarded as laws and natural kinds, and nothing which would choose between them. It is not that we do not know which is the real set of laws and kinds. Rather we do not have any notion of what it means to say that one is the real set and not the others.

It might be tempting to conclude that our concepts of law and kind really are empty, or nearly so, and that we are operating under an illusion if we think of all the things we call copper as sharing some property, or all so-called fruits as being similar. A slightly weaker but more justifiable conclusion is that there are laws and kinds, but that there is a high degree of arbitrariness about them. In this view the choice of our kind concepts are essentially conventional and not natural. Our choice is governed by our interests, not by nature. So there is no real difference between the classification of carrots as fruit or not. We can choose either classification, and which we do will be answerable only to the purposes of the classification. Neither is more natural. It might be thought that we conceive of some classifications as being more natural than others; we are able to see similarities and dissimilarities. Yet it is plausible that our perceptions of naturalness and similarity are interest relative and culturally determined. Anthropologists have shown that other civilizations find it natural to make classifications that strike us as bizarre.

This conclusion about laws might be welcome to some kinds of minimalist. For them the very idea of law is just that of generalizing about particular facts. For them what is really there are the facts, not the laws. So any classification or drawing up of regularities is as good as any other, so long as the required degree of simple, strong systematization is achieved. The minimalist is quite happy to regard laws and kinds as a structure imposed by us upon a reality consisting only of particular facts. By contrast, the full-blooded theorist wants none of this. For such theorists, laws, if there are any, are features of reality, not simply our interpretation of reality, and they have an important role to play in permitting explanation, induction, and prediction. So, for the full-blooded theorist the thought that laws and kinds are essentially arbitrary, conventional, or socially determined is unacceptable.

I think that this unacceptable conclusion is avoidable. As I remarked a few paragraphs back, conceptual circularity can be broken by giving concrete instances of a concept. (We saw that the use of samples avoids the problem that natural properties must be defined in terms of properties that are defined in terms of other properties, and so on.) To be able to give instances of natural kinds, and to some extent of natural laws, is an innate human capacity. We recognize that certain animals or plants belong to the same species and are different from others. We have an intuitive grasp of the law of gravity. We are able to recognize many cases of cause and effect.

That we have such a capacity should not, in one sense, be surprising, because such capacities have a survival value. If there are natural kinds and laws, creatures that are able to recognize them have a selective advantage over those that do not, for such animals will be able to recognize predator species and differentiate them from their own prey; they will be able to tell nourishing plants from poisonous ones; and so on. So Darwinism suggests that we should have some native ability to detect natural kinds.

The idea is that our concepts of natural kind and natural law are fixed by their internal relations plus their relation to certain intuitive natural kinds that provide basic instances. These instances form the foundations of science. It is by observing relations among these basic kinds that we discover our first unintuited natural laws. By speculating on explanations of phenomena we hypothesize the existence of new natural kinds. These new kinds will play a part in forming theories about further laws and kinds, and so on.

This edifice of natural kinds and laws is built upon the assumption of an innate ability to detect natural kinds. I mentioned a Darwinian argument which suggests that this assumption is plausible. But this argument is not a very strong one, given that there are many ways in which humans are imperfectly attuned to their environment – our intuitions about similarity might be a sort of cognitive appendix (vestigial, decayed and more likely to cause trouble than be of any use). Might we have started off with a completely different set of intuitions about similarity and constructed thereupon a completely different set of laws and kinds? A stronger argument is to be found in the reflexivity of science, by which I mean the readiness of science to subject its assumptions, methods, and techniques to scientific scrutiny. I will be looking at this idea in greater detail in a later chapter, but its relevance here is this. While science may have started out taking most of our intuitions for granted, it is now able to question those intuitions, justifying them where it can and rejecting and replacing them where necessary. Take for instance our discrimination of colour. This provides a very basic ground for classifying things, one that has played its part in the development of science, especially in chemistry and biology. It would be disturbing therefore if we discovered that there is no explanation of why we should classify red things together. Thankfully there is, as non-trivial discoveries in physics, psychology, and physiology tell us. In particular, we now know that colour is a function, among other things, of chemical structure, and so colour is a reasonable way of making distinctions between chemical substances. Similar remarks can be made about taste and smell, sadly neglected senses which nonetheless have been much used by chemists and biologists,[29] while the same goes also for our ability to make judgements of temperature and weight.

"So what?" a sceptic might respond. It is not especially convincing to find science justifying its own starting point. This looks like science trying to pull itself up by its own bootstraps. This charge I do not think is fair. First, the support lent by science to our intuitions is not simplistic. Rather it is sophisticated and critical. In some cases our intuitions are corrected by science. Natural as it is to think of whales as fish, we are now told that this is mistaken. Jade is two substances not one, though we might not immediately think so. Contrary to appearances, the Earth is not flat and stationary, but is spherical and rotates about the Sun. One instance of the detection of a limitation in perceptual capacities was the identification of colour-blindness in 1794 by the great chemist John Dalton. The fact that Dalton himself suffered from colour blindness illustrates that the investigation (and so also justification) of our innate capacities by science is not trivially circular, but instead is self-critical. Secondly, the explanation of colour perception uses parts of physics that do not historically depend much upon the use of colour discrimination. The physics in question could have been developed by a society of colour-blind scientists. So what justifies the use of colour discrimination is not itself a product of that ability. It is true that the various parts and branches of science are not entirely independent of one another. On the contrary, science is a complex web of theories that mutually support one another, in that the evidence for one theory might be its explanatory success when combined with another theory. The question to put to the sceptic is whether a distinct but similarly complex and successful network could both start from and justify a completely different set of intuitions about similarity. It should be remembered that our intuitions, though the starting point for science, are not its foundations. It is not that if we remove some of them the whole structure is in danger of collapse. A better analogy would be that of a seed crystal in a solution. The seed crystal is necessary for the growth of a crystal from the solution. Furthermore, it needs to be of the right sort; not any seed will precipitate crystallization. But once the crystal has formed around the seed, the seed may itself play little part in the structural integrity of the whole.

In summary, Goodman's problem shows that we cannot find a model of inference that allows us to decide, in isolation, which inferences it is reasonable to adopt. This should not surprise us; I have argued in the preceding chapters that there is no model for lawhood, explanation, or confirmation. In each case what it will be reasonable to regard as a law, explanation, or confirming evidence will depend on what other things it is reasonable to regard as such. The approach is holistic. What counts as a reasonable inference is in the same category, as are natural kinds. All these concepts are linked. Laws connect natural kinds. Induction can be seen as inference to the best explanation, where the explanation is a law. Therefore natural kinds are the kinds one should make use of in inductive inference. This interrelation among concepts raises chicken-and-egg questions. Is an inference reasonable because it employs natural kinds? Or is a choice of kind natural because it appears in reasonable inferences? As in the case of chickens and eggs, the answer is that the two evolve together. We have natural propensities to make inductive inferences with intuitively recognized kinds. Although these are not the basis of science, they are its starting point. Science itself reassures us that such a starting point is a satisfactory, if imperfect, one, and that our innate dispositions do by and large latch onto kinds identified as such by the esoteric discoveries of chemistry, mineralogy, biology, and so on. Perhaps you may find the validation that science gives to its own starting point is not so very reassuring. Is that not circular? That science does evaluate – *critically* evaluate – its own methods, is not to be doubted. Why this can provide genuine reassurance is another story, and is the topic of a later chapter.

Further reading

The Kripke–Putnam view of natural kinds is to be found in Hilary Putnam "The meaning of 'meaning' " and Saul Kripke *Naming and necessity*. This view is criticized in Eddie Zemach's "Putnam's theory on the reference of substance terms" and Hugh Mellor's "Natural kinds". W.V. Quine's "Natural kinds" presents an epistemology of kinds which has affinities to that promoted in this book. John Dupré's "Natural kinds and biological taxa" argues that the concept of a natural kind runs into difficulties when applied to biology.

Chapter 4

Realism

Realism and its critics

Eighteenth century chemists recognized a remarkable fact of chemistry, that elements react together in fixed proportions. We can mix oxygen and hydrogen in any proportion we care to, but in getting them to react to form water, they will do so only in the proportion of seven to one by weight, or one to two by volume. Other elements are found combined with one another in more than just one proportion, for instance carbon and hydrogen as methane and ethane, and oxygen and nitrogen as nitrous oxide and nitric oxide. Nonetheless, the proportions were always fixed and varying intermediate proportions were not to be found. In 1808, John Dalton resuscitated the atomic hypothesis in chemistry to explain this phenomenon. Dalton proposed that associated with each chemical element is a kind of atom. When chemical elements react together to form chemical compounds, what is happening is that the atoms of the elements are combining together to form clusters of atoms (molecules). If compounds are composed of identical clusters of this sort, then one would expect the elements to react together in a certain proportion, the same proportion indeed as the total masses of the different kinds of atom in the molecule. Nature might in some cases allow different kinds of molecules made up of the same elements and this would explain the case of the oxides of nitrogen.

However, Dalton's hypothesis was not universally accepted. Among scientifically minded philosophers Auguste Comte maintained in his "positive philosophy" that the business of science is the discovery of the laws governing the phenomena, which he regarded in a minimalist manner as correlations among observable qualities and quantities. By "reasoning and observation" the scientist will be able to establish "invariable relations of succession and resemblance" seeking as far as possible to simplify and generalize them, for instance by reducing particular laws to more universal ones. As for causes and hypotheses about hidden entities and mechanisms, Comte regarded these as vestiges of a metaphysical state of thought not far removed from theology:

> What scientific use can there be in fantastic notions about fluids and imaginary ethers, which are to account for phenomena of heat, light, electricity and magnetism? Such a mixture of facts and dreams can only vitiate the essential ideas of physics, cause endless controversy, and involve the science itself in the disgust which the wise must feel at such proceedings.[30]

As regards the atomic hypothesis, Comte acknowledged its usefulness, but, he maintained, the important thing is the truth (or otherwise) of the law of proportions. Once that was established by careful measurements, the question of atoms could safely be ignored.

The great chemist Sir Benjamin Brodie shared Comte's "positivist" approach. He argued that all a chemist could ever see is the reacting together of certain quantities of substances consequent on actions performed by the chemist. Since this is all a chemist is ever in a position to observe, it is to this that a chemist should confine his reasoning. The chemist had no business extending his speculation to the unobservable realm of the constitution of matter. Brodie instead devised an ingenious if elaborate system whereby different substances were regarded as the result of chemical operations performed on a unit volume of space. Chemical equations thus became algebraic equations, which could be solved without any consideration of the unobservable properties of the substances in question. (Ironically, Brodie accepted only whole-number solutions to his equations, which seems an arbitrary requirement, unless one accepts the atomic point of view.[31]) Brodie's contemporaries, unable to fathom the complexities of his chemical algebra, were more ready to use Dalton's hypothesis as a heuristic model, that is to say as a useful way of capturing the fact that elements react in fixed proportions. Nonetheless, they agreed with Brodie that nothing a chemist could observe could be regarded as supporting a belief in the physical existence of atoms.

The debate over the atomic hypothesis illustrates opposing attitudes prevalent in the philosophy of science. Dalton regarded his hypothesis as a claim that might well be true, for which evidence could be found, the truth of which consisted in the existence of unobserved entities (atoms), entities which would explain the observable phenomena (the law of fixed proportions). Against this realist view of science, a typical anti-realist argues that science should aim at no more than the ordering and correlation of the phenomena. However useful the hypothesis is in this respect, we should not think of it as being literally true. Since we should not think of atoms as genuine entities we should not think of the hypothesis as explaining the law of fixed proportions. Rather it is another way of expressing the law of fixed proportions. To see Dalton's hypothesis from a realist perspective, as providing an explanation of the phenomena, and to see the fact that it would provide such an explanation as evidence in favour of its being true, is, in the eyes of the anti-realist, unscientific speculative metaphysics.

We have already seen manifestations of this difference of approach between those who want to minimize the metaphysical commitments of science and those who are happier with a much richer metaphysical diet. The full-blooded theorist wanted laws to be facts that explain observed correlations, and so was happy to employ such notions as necessitation among universals. The minimalist regarded nomic necessity as going too far beyond what could be observed. If nomic necessity goes beyond observation its existence cannot be verified by observation. Worse, if the relevant concept goes beyond what can be understood in terms of what is available to sensory experience, then that concept (e.g. "nomic necessity") is in fact meaningless. Whether meaningless or unverifiable, nomic necessity is no business of science. Similarly, a philosopher with an anti-realist bent may regard the thought that there are natural kinds the distinct existences of which are natural facts as a metaphysically overloaded fancy. The most we can say is that there are certain classifications which enable useful correlations to be constructed. In this chapter we will look at anti-realism in some of its most explicit and general forms. The realist, like Dalton, wants to say of his theories that:

(a) they can be evaluated in terms of their truth or nearness to the truth;
(b) they reasonably aim at truth or nearness to the truth;
(c) their success is evidence in favour of their being true;
(d) if true, the unobservable entities they hypothesize would genuinely exist; and
(e) if true, they would explain the observable phenomena.

Scientific anti-realism comes in a variety of forms, depending upon which of the above is denied. If an anti-realist denies (a) then (b)–(e) are simply irrelevant statements. (b)–(e) are premised on the possibility that a theory can be thought of as being true, a premise which some anti-realists regard as being mistaken. This is the most revisionary of the anti-realist positions, since it rejects what seems a most natural approach to theories, that they may be true or false depending on the way the world is. In this view Dalton's hypothesis is neither true nor false. We shall presently see how this view may be spelt out in detail, and how it fares.

A more subtle version of anti-realism avoids this unnaturalness by accepting (a), but denying that (a) entails (d) and (e). This anti-realist accepts that theories may be true, but argues that if we examine carefully what the truth of a theory amounts to, we will see that the truth of a theory does not involve the existence of unobservable entities or its having explanatory power. Accordingly, Dalton's hypothesis should be accepted as (potentially) true, but to think of it as being true is not to think of there really being unobservable atoms the existence of which explains the observable facts. This version of anti-realism is less revisionary than the first, but still requires us to change some of our pre-philosophical attitudes towards theories, in that we may not take their meaning at face value.

Lastly, we shall look at that species of anti-realism which accepts (a), (d), and (e), but denies (b) and (c). This anti-realist, the *constructive empiricist*, goes along with the natural thought that theories are either true or false, may be nearer or further from the truth, that if true entail the existence of unobservable entities, and would have certain explanatory force if true. What the constructive empiricist denies is that this has any relevance for science. Science cannot reasonably aim at truth. It should aim at success, but the inference from success to truth or even nearness to the truth is too weak to be of any value.

Instrumentalism

Instrumentalism is so called because it regards theories not as attempts to describe or explain the world but as instruments for making predictions. For the instrumentalist, asking about the truth of a theory is a conceptual mistake. The appropriate question to ask is whether a theory is *empirically adequate*. A theory is empirically adequate if it makes accurate predictions, i.e. if all its observable consequences are true. The black-box analogy is an appropriate one for instrumentalism – the theory is like a black box (Fig. 4.1). One puts into the box information regarding observed background conditions, and the box generates predictions regarding what one will observe. What one wants from such a black box is that if the input information is accurate then the predictions

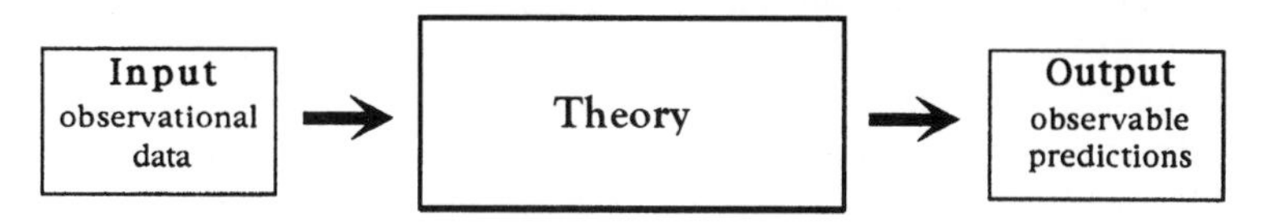

Figure 4.1

it yields will be accurate too. We are not especially concerned with the mechanism inside the box. That can be anything so long as it works. In particular, there is no requirement that it depict the way the world is. Indeed, it does not make sense of the mechanism to say that it does or does not depict the world – it is merely an instrument.

If a statement or some other thing, like a proposition or belief, is something of which it makes sense to say that it is true or that it is false, then we say in the jargon that it is *truth-evaluable* (or *truth-apt*). So assertions like "the solution is coloured red" is truth-evaluable, while other sorts of utterance such as questions and commands are typically not truth-evaluable. According to this form of instrumentalism theories are not truth-evaluable.

While the theory itself is not truth-evaluable, it is clear that the information put in and the predictions produced are truth-evaluable. The instrumentalist is therefore committed to the existence of two kinds of statement in science: theoretical statements and non-theoretical statements, where the latter are truth-evaluable and the former are not. Correspondingly, there are two kinds of term: those that refer to things which exist and those that do not refer to things which exist. Essentially there are two sorts of language: that which is not theoretical and that which is.

The principle that demarcates the non-theoretical from the theoretical is *observability*. For a statement to be truth-evaluable it must be such that its truth can be determined by observation. For a term to be a referring expression (one that refers to or can refer to existing entities or properties) it must have its meaning fixed by its being associated with some observable entity or property. Thus the truth-evaluable statements are called O-statements and the referring expressions are called O-terms. Scientific statements that are not O-statements and occur in theories are called T-statements, and non-referring expressions are called T-terms. An O-statement cannot contain any T-terms. ("T" here stands for "theoretical".) Mixed statements, containing both O-terms and T-terms will be classed as T-statements. So statements like "the voltmeter needle points to 9" and "the paper is blue" are O-statements. Statements like "the electro-motive-force is 9 V" and "the solution is basic" are T-statements, as are the mixed "the falling reading on the electroscope is the result of bombardment by cosmic muons" and "red litmus paper indicates a concentration of H_3O^+ ions of greater than 10^{-7} mol l^{-1}".

This definition of an O-statement needs some qualification. Let it be agreed for the sake of argument that the statement "this is a strawberry and it is red" is one which can have its truth or falsity determined by observation. The question is what to do with a statement such as "all strawberries are red". It might not be possible to determine the truth of this statement, were it true, directly by observation since it might not be possible to observe all strawberries (especially if we are concerned with future strawberries, or ones which have just been eaten by wild animals. And would we anyway know we had seen them all?). So we need to say something about the meaning of general statements. In this case the meaning is composed of two elements: that for some unspecified *x*, if *x* is a strawberry it is red, and that the latter is true for all *x*. Once *x* has been specified, and so the first half becomes a determinate proposition, then the truth of that proposition becomes an observable matter, at least in principle. The second half is a logical operation, generalization. So one might extend the notion of an O-statement to include those statements that, although not open to having their truth determined by observation directly, are constructed out of O-statements (and only O-statements) using purely logical operations, i.e. consist only of O-terms and logical terms.

So we have two sorts of statement: those the truth of which can be determined by observation or are logical functions of such statements, and those that contain theoretical terms. The meaning of the former has been described, but what of the latter? These have meaning in a quite different sense. The meaning of O-statements is connected with the observable conditions under which they are true, but T-statements cannot be true, nor can they be false. Their role is to be part of the mechanism within the black box. Correspondingly, their meaning is understood in relation to the theory as a whole. The meaning of T-terms is not given by association with observable things or properties, but is to be understood in terms of their systematic role within the theory. For instance, above I wrote of "muons", "bases", and "electromotive force." These terms cannot be explained by pointing to something we can see. Rather, to explain them we must explain various theories of particle physics, of acids, of electrochemistry, and of electrodynamics. According to this view, to understand the term is to understand the whole theory of which it is a part and to see how it fits into that whole. The two kinds of statement and the two kinds of term mean we have two sorts of language: O-language and T-language. Their differences can be summarized as shown in Figure 4.2.

There is an alternative view within the instrumentalist camp that concerns the meaning of T-statements and which needs mentioning. This is that T-statements are genuinely truth-evaluable. They are truth-evaluable because they are equivalent to some complicated function of O-statements. The idea is still that proper meaning resides with the O-statements. The role of theory is to transform data O-statements into prediction O-statements. So the theory as a whole can be seen as a large conjunction of conditionals:[32] *if* observations of this kind have been made *then* these observations will be made, *and if* observations of that kind have been made *then* those observations will be made, *and* so on. This big conditional is just a truth function of O-statements. One might think that it is possible, in principle, to analyze each T-statement into a similar truth function of O-statements.

Although this was the preferred view of T-statements among some positivists, it has its disadvantages, even when compared with the view first outlined. It is not at all clear how one could

O-language	*T-language*
• *O-terms have their meaning fixed by correlation with things or properties*	• *The meaning of T-terms is determined by their systematic role within the theory*
• *O-terms refer*	• *T-terms do not refer*
• *O-statements have their meanings fixed by observable truth-conditions*	• *The meaning of T-statements is a function of their role within the theory*
• *O-statements are truth-evaluable*	• *T-statements are not truth-evaluable*

Figure 4.2

generate the equivalent (truth function of) O-statements for individual T-statements, nor indeed whether there really is such an equivalent. The problem is that T-statements do not function individually but within the theory as a whole. A solitary T-statement does not on its own link data and predictions but only when combined with other T-statements. (See pp. 174–6 for an explanation of why this is.) The process of finding an equivalent looks very difficult. Perhaps it is impossible. (And, if it is possible, why should there be precisely one translation into O-language?)

The matter is further complicated when we consider that what has just been said about the interdependence of T-statements also goes for whole theories as well. Theories typically do not connect with observation just on their own. More often theories will be combined to make predictions. So a theory about the composition of a distant star will need to combine with the theory about red shift to give predictions about its observed spectrum.

This means that the black box connecting O-data with O-predictions consists not of single theories but of the whole of science. Thus the process of analysis will be exceedingly complicated, and with no guarantee that there will be an analysis or, if there is one, that it will be unique.

A more important problem faces both forms of instrumentalism. Both require a sharp distinction between O-language and T-language. This is because these are semantically different. An O-statement is truth-evaluable, a T-statement is not. There is no in-between state of being partly truth-evaluable. Similarly an O-term refers to something or to a property, while a T-term does not. The idea that a term could partly refer to something but not completely refer to it is absurd. So the distinction between O and T must be sharp.

However, it appears that the distinction between observation and theory is not at all sharp. The boundary is, if anything, vague and context dependent. Indeed, it may not be appropriate to talk of a boundary at all, since theory and observation concern different kinds of things, which while they often exclude one another may also coincide, or even both be absent.

This is because the distinction between theoretical and non-theoretical is *semantic* and the distinction between observational and non-observational is *epistemic*. The theoretical/non-theoretical distinction is semantic because it involves the different ways in which an expression can get its meaning. A theoretical term gets its meaning from the role it plays within a theory, while a non-theoretical term does not – for instance it may get its meaning from ostensive definition (ostensive definitions are those involving a sample, e.g. one might define "pink" by pointing at a carnation). The observational/non-observational distinction is epistemic, because whether we can observe something or not is a question about how we can know about it. The instrumentalist treats these distinctions as if they are the same (In reverse, so theoretical = Non-observational and observational = Non-theoretical). But the semantic and epistemic distinctions appear to be quite different. It may well be that the meaning of the term X is given by a theory, but that Xs themselves are perfectly observable. The terms "gene" and "supernova" are both expressions that get their meaning from a role played within theory. Yet genes and supernovas are both things that are easily observable. At the same time there are terms that are not semantically theoretical, and yet are not names of observable entities. Names of abstract entities are perhaps like this: "the Greenwich meridian", "the positive square root of nine". Other cases may involve entities that may be observable in principle but are not observable in practice (e.g. the star that is the furthest from the Earth).

Instrumentalists are inclined to think that the distinctions are the same because they think that the basic way in which words get their meaning is by ostensive definition. This is how non-theoretical terms are defined, e.g. "pink" by using a pink carnation as a sample. But for this to be possible the kind of thing being defined must be observable (in this case the colour of the flower). It is then thought that theoretical expressions refer to things that cannot be defined in this way, i.e. non-observational things. But it seems wrong that theoretical definitions are just a way of defining things that cannot be defined ostensively. A theoretical definition can provide information about the semantics of a word which cannot be supplied by an ostensive definition. For instance, footprints are visible things, but they cannot usually be defined ostensively because part of what is meant by saying that something is a footprint is not just that it is a foot-shaped mark but also that it was actually caused by a foot. So footprints are both observational

	Observational	Non-observational
Theoretical	e.g. gene, supernova	e.g. neutrino
Non-theoretical	e.g. the colour pink	e.g. the furthest star

Figure 4.3

and, in a small way, theoretical. But the case is similar for genes and supernovas, as their semantics include information about their causal roles, origins, etc. Hence their definition is not ostensive but theoretical. Looking at the grid in Figure 4.3 the instrumentalist thinks that there should be entries only in the bottom left and top right boxes and none in the other two.

As the observational/non-observational distinction is an epistemic distinction, it is difficult to see how it can support a semantic distinction. The epistemic distinction is vague. But the semantic one cannot be. As we saw above the distinction between O-language and T-language is sharp. O-statements are truth-evaluable but T-statements are not; O-terms are referring but T-ones are not. A statement cannot be partly truth-evaluable; a term cannot refer to something to some extent but not fully. Either a statement is truth-evaluable or it is not; a term is either referring or it is not. According to the instrumentalist, what makes the difference between the O-language and the T-language is observability, and so the distinction between the observable and the unobservable needs to be correspondingly sharp.

Observation

Let us look more carefully at what is meant by observation. Given the connection with ostensive definition, instrumentalists tend to mean "discernible by unaided sensory experience" when they say "observable". It is worth remarking that scientists do not use the word in that way. Observations are typically made using instruments (telescopes, microscopes, Geiger counters, spectroscopes, electroscopes). The very fact that so many of these devices are called "scopes" is itself suggestive. The suffix comes from the Greek word *scopeo* (σκοπεω), which means to consider, examine, watch, or observe. The fact that one is using an instrument, far from precluding one from observing, enables it. A scientist using a cloud chamber to make observations will report what they see by saying "The creation of an electron–positron pair was observed", and not by saying "A spiral to the left and a spiral to the right originating at the same point were seen". It would, however, be less appropriate to say of someone who knew nothing of physics or of cloud chambers that they saw that an electron–positron pair was created. Observation is a matter of training, it is a matter of recognizing what one is looking at.

Some philosophers have considered this fact, that two people can look at the same thing and one be said to be observing electrons and the other not, and have concluded that there is nothing objective about observation. I think that such philosophers have thought that if observation is to be objective it must be a matter of the objective world impressing itself upon the passively receptive mind. Thus two similarly passive receptive minds ought to observe the same thing in the same circumstances. But this is mistaken. Observation is not passive. It is an activity. Like most activities it is one which can be performed more or less well. It is an activity with a purpose, which is the production of reports or beliefs about the world. More specifically, it is supposed to produce *reliable* beliefs and reports. The importance of reliability is that observation is not an end in itself. Rather observations are used in the testing of hypotheses. In order that our testing procedures should themselves be reliable, and so our beliefs in the theories thus generated be reliable, it is important that the observations be reliable. Indeed, this I think is a necessary feature of being an observation – that it is a belief or report produced by a reliable process. The reliable process may involve very sophisticated instruments, constructed on the basis of complicated theory. But, as long as they are reliable, then it may be possible to use them in the making of observations. What goes for the instruments also goes for the observer. The observer needs to be able reliably to recognize the facts being observed. If the observer is not reliable then neither will his reports be reliable.

It is rather less clear whether it is *sufficient* for observation that a reliable investigator make a true report, using a reliable instrument – articulating full necessary and sufficient conditions for observation is a difficult, perhaps impossible task. Nonetheless, the core of the notion of observation I am proposing should be clear – to observe something is to be able to make reliable reports about it. It is clearly a very different notion of observability from that needed by the instrumentalist. It allows for the observability of highly theoretical items such as positrons. Whether a statement counts as an observation will depend upon the reliability of the observer and his instruments. Furthermore, this will allow for a change in status from observable to unobservable. As instruments and observers are developed, trained, and become reliable, so more things will become observable.

Hence whether a statement is an observation statement will not be fixed by the language in which it is couched, but will depend upon extra-linguistic matters. Thus whether or not a statement is an O-statement cannot be a semantic matter. Nor is the distinction between observable and unobservable a sharp, fixed one. Instruments and observers can increase in reliability, so that whether or not a statement counts as an observation may be a vague matter. Things can be more or less observable depending on the availability of instruments (and scientists) with the capacity to report on them reliably. So not only is the observation/non-observation distinction not a semantic distinction, it is not a sharp distinction either. Hence it cannot mark the sharp distinction needed between two semantically contrasting languages.

The foregoing remarks also serve to raise a more general question, i.e. whether any debate about anti-realism founded on an observable/unobservable distinction is really well-founded. This goes for both instrumentalism and constructive empiricism, which I will discuss next. The debate assumes that much of that which science is concerned with is unobservable. But is this really true? I remarked that with this notion of "observable" positrons are observable. This accords with the normal parlance of scientists according to which very little is "unobservable". Neutrinos used to be another good example of an unobservable entity. The only reason for thinking they must exist was to account for deficiencies in mass-energy and angular momentum after certain subatomic reactions (β decay of radium-210). Wolfgang Pauli hypothesized the existence of neutrinos in 1930, and in 1934 Enrico Fermi provided a quantitative theory of the processes in which they are involved, which accurately accounted for the half-lives of the relevant nuclei. Such was the success of the theory that physicists accepted the existence of these particles, which have zero rest mass and no charge, without any observation of them. Indeed, the theoretical nature of the neutrino was such that the idea of detecting it seemed an impossible one. Only one in 10 billion neutrinos passing through the Earth will cause a reaction, the others go through undisturbed. The average neutrino can travel for one light-year in lead without interacting.[33] So the neutrino would seem to be an unobservable entity par excellence. Nonetheless, Frederick Reines and Clyde Cowan, in a famous experiment performed in 1956, conclusively detected neutrinos emitted by a nuclear reactor in Georgia. This experiment involved a large tank of fluid with which some of the many emitted neutrinos interacted by being captured by protons. The products of this capture (the reverse of decay) could be seen with the aid of sensitive detection equipment. This is now regarded as a standard way of detecting neutrinos, although it is by no means easy to carry out.

The philosophical point of this example is that, if the scientists are to be taken at their word that they have observed neutrinos, then the large majority of entities discussed by science count as observable. If we are thus entitled to be realist about so much, then the debate about anti-realism is limited to that small and abstruse corner of physics that hypothesizes entities for which no successful detection experiments have been devised. Since scientists are ingenious (and in this area well-funded) they can, with the aid of ever more powerful accelerators and other equipment, detect more and more of what is hypothesized. So it might seem that anti-realism is a philosophy with a very limited application.

In the foregoing discussion it must be admitted that I have freely moved between talk of *detecting* particles and *observing* them. There is a difference between these two notions: one can detect an animal by its tracks, but this does not normally count as observing it. Whatever the difference between the concepts, they are clearly linked, and one of the features that links them is the fact that they are both success words. If one detects a neutrino there must be a neutrino there, and similarly if one observes the existence of a neutrino, the neutrino must exist (contrast with *hypothesizing* the existence of a neutrino). For that reason it would not help the anti-realist to deny that we observe neutrinos while admitting that we do detect them.

So the anti-realist ought to deny that experiments such as the Reines–Cowan experiment constitute either an observation or a detection of neutrinos. After all the experimental set-up in question involves a lot of theory, and the anti-realist may wish to deny the truth (as opposed to empirical adequacy) of such theories. I have said that I think that the notion of observation (and detection too) is bound up with the idea of reliability. In a later chapter I will say why this is right and, therefore, why it supports the claim that subatomic particles are observable or detectable. For the time being it should suffice to note that, according to the parlance of the physicist, neutrinos have been detected. So if the anti-realist denies this, he ought to explicate and defend a different notion of observability. I do not think this can be done. However, in what follows, I will give the anti-realists a run for their money and assume that it can be, and that "highly theoretical" entities are by and large unobservable.

Constructive empiricism

One form of anti-realism that is popular today is Bas van Fraassen's *constructive empiricism*. Constructive empiricism does not maintain a difference in kind between observable entities and unobservables, but does hold that because there are significant differences in empirical support for claims about observables and for claims about unobservables, we cannot have the same degree of belief in the two. Two new notions need to be introduced. Two theories are *empirically equivalent* if they have the same consequences about observable entities, properties, and relations ("observable consequences"). A theory is *empirically adequate* if its observable consequences are true.

The notion of empirical adequacy is important for understanding constructive empiricism. An important distinction needs to be made between truth and empirical adequacy. A true theory is necessarily empirically adequate. If it is true then all its consequences, including the observable ones, are true. But an empirically adequate theory is not necessarily true. Indeed, an empirically adequate theory chosen at random is unlikely to be true, although by remote chance it might be. This is because a lot of false theories can be empirically adequate; for any given true theory there will exist, even if as yet not thought of, false theories that are empirically equivalent to it. For instance, our theories about electrons and atomic structure have as their consequences the familiar phenomena of electrostatics, e.g. the ability of friction to make balloons stick to the wall. In seeing that sort of phenomenon we do not see any electrons – electrons are in that part of the theory which goes beyond empirical adequacy. That being the case, another theory could have precisely the same consequences about electrostatics without mentioning electrons at all. The idea is the black box of instrumentalism again. We know our black box works – it does the job required of it; in the case of a theory it delivers true propositions about the observable world. But there are many different internal designs of black box which could produce the same true output. These black boxes are all empirically equivalent and each is empirically adequate.

That being the case, if we are given a choice of black boxes, which one do we choose? Our aim may be to get a true theory, i.e. we want a black box the internal structure of which matches the structure of the world, and the claims of which about the existence and nature of unobservable entities are true. But we have nothing to go on in making a choice for truth. All we could have is the evidence, and since each theory is empirically adequate and equally so, the evidence is the same for each. Similarly, if we have already chosen a theory, because for historical reasons it is the preferred theory of the day, we should not believe that its claims about unobservables have a special credibility. This is because there exist in the abstract (although we may not have considered them) alternative theories that are equally empirically adequate, but which make quite different claims about unobservable entities.

The long and the short of it is that the evidence is the same for empirically equivalent theories. This is the case even if one of the theories is true and the other is false but empirically adequate. So the evidence gives us no reason to think that our favoured theories are true as opposed to merely empirically adequate.

An important feature of constructive empiricism is that it does not depend on the notions of "observable consequence" or of "evidence" having sharply defined boundaries. Nor, correspondingly, do "empirically equivalent" or "empirically adequate" need to be especially precise. We can make do with a vague notion of observable that shades off into the unobservable, so things can be more observable or more unobservable (some things will be very obviously observable, such as bicycles and lumps of stuff, and others very unobservable, such as neutrinos). So two distinct theories may be roughly empirically equivalent if they agree when it comes to the more observable end but disagree in their consequences at the unobservable end of the observable–unobservable spectrum. In the middle ground they may diverge only slightly. So theories A and B agree in their consequences in the more observable arena, diverge considerably with regard to the highly unobservable, and disagree mildly in the middling observable area. The anti-realist argument can still run, since a small amount of evidence of middling observability for B over A will not be very strong grounds for preferring B, and so will only be weak evidence in favour of B's claims with regard to the existence of middling observable entities. Furthermore, for more deeply unobservable entities we can run the argument again. For there will be a theory C similar to B with regard to consequences of high and middling observability, but diverging with regard to more deeply unobservable entities. Between B and C the evidence will not assist in making a choice. So the conclusion is that the more unobservable an entity is, the less evidence there is that we could have to believe in it and the further empirical adequacy diverges from truth.

I think that constructive empiricism is mistaken. I am going to develop two reasons for this. The first is a negative reason: if realism is untenable, then so is constructive empiricism. Roughly speaking, the argument is that if we cannot know theories to be true then we cannot also know them to be empirically adequate. The second argument takes issue with the constructive empiricist's notion of evidence. An appropriate notion of evidence does provide support for realist claims. I will conclude by looking at anti-realist arguments against Inference to the Best Explanation.

Laws and anti-realism

So far we have been looking primarily at anti-realism with regard to unobservable entities and the theories that purport to mention them. However, this need not be the only target or motivation for anti-realism, and another anti-realism, one which is more plausible to my mind, attaches to theories concerning what laws there are. So, before looking at criticisms of van Fraassen's version of anti-realism, I want to look at a different version that may not be subject to all of the same objections.

Consider the various theories of atomic structure that have been developed since the late nineteenth century. All agreed that atoms must contain a small negatively charged particle, the electron, although the theories differed radically as to the structure of the atom and the nature of the laws governing its components. One might reasonably think that this agreement on the existence of the electron establishes that the electron is quite solidly a fact, while the disagreement on laws (first classical mechanics then quantum mechanics) means that the facts as regards the latter are in greater doubt.

Think also about the laws of motion. The success of Newton's mechanics may once have been thought to demonstrate its truth. But it is now known to be false and has been superseded by Einstein's relativistic mechanics. What is instructive about this case is that the agreement between Newton and Einstein is very close for objects travelling at low velocities, the theories diverging markedly only for objects travelling at velocities close to the speed of light. This suggests two things. First that our theories as regards the laws of nature may be radically wrong for circumstances quite unlike those we have experienced. And, secondly, even in circumstances of an every-day sort our proposed laws may be wrong by an amount too small for us to detect. Most functional, i.e. quantitative, laws involve constants. So Schrödinger's equation involves Planck's constant *h,* and the law of gravity requires the gravitational constant *G,* and so on. One of the tasks of science involves measuring these constants to as high a degree of accuracy as possible. But we could always, in principle, measure more accurately. Furthermore, it is conceivable that some of the constants are not constant at all, but are changing very slowly over the history of the universe in accordance with some as yet undiscovered law. So at best our theories as regards laws will only be approximations (though perhaps highly accurate approximations); at worst, these supposed laws are only approximations for those circumstances and portions of the universe and its history with which we are familiar, while the real laws diverge significantly in circumstances, places, and eras with which we are unfamiliar.

Nancy Cartwright[34] has further arguments that aim to show that we should not believe our theories of fundamental explanatory laws. She distinguishes these from phenomenological laws. The latter concern regularities that can be established on the basis of observation. The realist thinks that we can also establish the existence of fundamental laws that explain the phenomenological ones. The realist has in mind Kepler's phenomenological laws, which describe the motions of the planets on the basis of careful observation. These are in turn explained by Newton's fundamental laws of gravity and motion, which also explain Galileo's laws of falling bodies, and so on. Our belief in Newton's laws is licensed by inference to the best explanation. The realist is inclined to think, as the example suggests, that while the phenomenological laws are many the fundamental ones are fewer. He or she may also think that our knowledge of the latter is more accurate – so in these cases while Gm_1m_2/r^2 accurately describes the gravitational forces between things, Kepler's and Galileo's laws do not describe their motions precisely.[35]

In many cases, however, the matter is reversed. For example, the behaviour of lasers is well tested and documented – the phenomenological laws are well established. But the quantum theory that seeks to explain such phenomena is, in many regards, a mess and not well understood at all. Furthermore, insofar as our fundamental theories are accurate, they are no good at explaining, and insofar as they are useful for explaining they are not true.

Take the equation for the law of gravity just mentioned. I used it to explain Kepler's and Galileo's laws. But is it true? When is the force operating on two bodies exactly Gm_1m_2/r^2? Only when no other forces are operating. For if there are other bodies around, or if these bodies are themselves charged, or if they are particles in a nucleus, then other forces will be operating and the forces experienced by these bodies may be quite different. So the equation describes the forces that exist only when none of these other things occur – which is never. We can regard the equation for the law of gravity as true only if it is limited to non-existent circumstances, in which case it cannot explain anything, such as planetary motion, which does exist. So to explain the latter, the equation must be regarded as, at best, an approximation that is strictly false.

Of course, there are ways a realist can respond to this attack. One is to suggest that what the gravitational equation supplies is not the force between two objects but something else, a causal influence.[36] Distinct influences are supplied by the different equations (for the gravitational, electrostatic, magnetic, strong, and weak nuclear forces[37]). A further law, a law of vector addition, details how these combine to produce the appropriate motion in the bodies concerned. Cartwright has various responses to this move. One is that it cannot be used in general, and so lacks the universality we would expect from a law operating at the fundamental level; secondly, these causal influences cannot be identified experimentally – they seem to Cartwright to be "shadow occurrences which stand in for the effects we would like to see but cannot in fact find".[38]

We cannot follow this interesting but difficult discussion further. Nonetheless, it serves to show that the issues arising from anti-realism about (our knowledge of) laws can be regarded as distinct from those concerning entities. In addition, two further things are worth mentioning. First, the objection to causal influences demonstrates a typical anti-realist resistance to things that are not, it is claimed, experimentally detectable (although a realist may debate this). Secondly, Cartwright takes the gravity equation to be explanatory only when it is false. I have emphasized that only something that exists can properly explain. So the notion of explanation with which Cartwright is working must be very different from mine (as indeed it is). In fact, it is characteristic of anti-realists that they take explanation to be achieved not by facts (i.e. laws or causes) but by other things: theories, models, and law statements.

The success of science

In this section I will look at an argument which, while very popular, is not one I myself find convincing (in the remaining sections I will present what I take to be better arguments against anti-realism). This argument is best summed up in an oft-quoted remark from Hilary Putnam: *"realism ... is the only philosophy of science which does not make the success of science a miracle".*[39] Accordingly, realism is seen as a theory, for which the success of science is the evidence: science is successful and this is best explained by science being, by and large, true and it genuinely referring to existing unobservable entities. Thus our belief in realism is itself justified as a case of Inference to the Best Explanation.

As the term is being used by Putnam, realism is certainly an empirical claim. It is an empirical matter whether theories are true and whether the entities they mention exist. It is less clear that realism should be thought of as a theory, and if it is a theory whether it is a theory distinct from the theories collected together under the heading "science". The reason why one might doubt that realism is a theory is a doubt whether realism is a genuinely explanatory claim. Let us focus on the truth part of realism. Roughly speaking, this says that "the theories of science are (by and large) true"; this is supposed to explain a phenomenon, the success of science. This is an odd-looking explanation. It regards truth as a property, and a property with explanatory force. This seems implausible; if truth is any sort of property then it is a property of propositions or statements, or sets of such things, such as theories. It is unlikely that there are any laws of nature that make reference to propositions, and so it is unlikely that there are any genuine explanations involving them or the property *truth* which they may possess. (Perhaps there are sociological laws or psychological laws governing languages and theories and so truth as well, but the realist does not want it to be these that explain the success of science.)

The nature of truth itself is a contentious philosophical question and one which it is beyond the scope of this book to examine in detail. In any case, it seems that, on any understanding of truth, it cannot have the explanatory role the realist would like it to have. One extreme, the minimalist or deflationary view of truth, regards truth as an almost contentless notion – certainly no property. To say that a statement S is true is tantamount to merely repeating S itself. "It is true that the Sun's energy comes from nuclear fusion" is equivalent to "The Sun's energy comes from nuclear fusion". At the other extreme among accounts of truth, the correspondence theory says that truth is a genuine property of propositions, which marks their correspondence with the world. On the former view, truth cannot explain anything, since it is no property. On the correspondence view, truth is a property. But it is a property a proposition has solely by virtue of the existence of a certain fact plus the correspondence of the proposition to that fact. So if a proposition *p* has any explanatory power over and above the explanatory power of the corresponding fact that *p*, then that additional power must be by virtue of its linguistic structure and meaning. But such things are unlikely to contribute much to the success of science. So if the truth of S explains anything, then this will be by virtue of the fact that corresponds to S. Does the *truth* of the sentence "The Sun's energy comes from nuclear fusion" explain anything not also explained by the *fact* that the Sun's energy comes from nuclear fusion?

If we turn to the explanandum, the success of science, then we can say similar things. The success of science means the corroboration of its predictions and the failure to falsify them. This, in turn, for reasons similar to those just given, is tantamount to the existence of certain facts and the non-existence of other facts. And so, if it is right to say that the truth of science explains its success, this is to say no more than the facts hypothesized by the theories of science explain the facts which are the evidence for those theories.

Consequently, it appears to me that the argument that the truth of science best explains its success is no more than an amalgamation of the various arguments for each of our best theories that they best explain the evidence pertinent to them. There is no additional argument for realism as a hypothesis than the reasons we have for believing quantum electrodynamics, our reasons for believing in plate tectonics, our reasons for believing in the ring structure of benzene, and so on.

This makes realism into something like a default position. If we regard the evidence for plate tectonics as justifying belief in it, then we are realists, at least as far as plate tectonics is concerned. It should be added, of course, that if we do not think our evidence for our best theory is very strong, we may take a weaker attitude, e.g. accepting it as a satisfactory working hypothesis. What makes constructive empiricism interesting is that it argues that we should always take a weaker attitude, that there are quite general considerations why we should never regard our evidence as strong enough for belief. Van Fraassen thinks that the appropriate attitude to take towards a successful theory is one he calls *acceptance*. To accept a theory is not to believe it, but is instead to believe it to be empirically adequate (there are other aspects to acceptance, e.g. engagement in a research programme and using a conceptual scheme). What I have been suggesting above is that the Putnam argument is just a generalization of the claim that each theory is well supported by its evidence. As it provides no additional argument, it has no force against the constructive empiricist thinking that our best theories are not as well supported by their evidence as we the realists supposed.

There are two conclusions we should draw from this. The first is that we cannot buy Putnam's "truth explains success" argument against anti-realism. On one understanding of "truth" it is no argument at all, and even on the stronger reading of "truth" it is no argument against the anti-realist. The second conclusion is that we should not expect, nor do we need, a strong argument *for* realism. Realism, in Putnam's sense, when addressed to a particular theory is just the view that it is well supported. Which, in the absence of arguments to the contrary, is a matter usually best judged by the scientists themselves. So what the realist should do is not advance arguments on their own behalf, but try to refute the general arguments of the anti-realist, which is what I shall go on to do.

Anti-realism and inference

The problem for constructive empiricism, and anti-realism in general, which I raise here concerns inference. If science is to work, it must involve inductive inference or something like it. (Popper thinks otherwise, and in a later chapter I will discuss his views.) The question for constructive empiricism is this: How can it account for scientific inference? The problem is that the constructive empiricist must give an account of inference (or, equally, of evidence) such that what we infer from the evidence is not the truth of a theory but its empirical adequacy. I do not think this is possible. Our inferential practices do not work if we regard them as inferences to empirical adequacy and not to truth.

In Chapters 1 and 2 I explained why I think that many, perhaps all, our inferences are instances of or employ Inference to the Best Explanation. Even simple induction shows this. I argued that the inference from "all observed lumps of iron are attracted by magnets" to "it is a law that all lumps of iron are attracted by magnets" should be understood as an Inference to the Best Explanation. What happens in Inference to the Best Explanation is that we infer from the evidence that such and such is the explanation of that evidence – that the law explains the magnetic behaviour of pieces of iron. The explanation consists of facts – in this case the fact is that law. Those facts are thereby what is inferred.

The question is, what attitude should the anti-realist take towards Inference to the Best Explanation? Clearly the anti-realist rejects explanation as a guide to truth. Should explanation be taken to be a guide to empirical adequacy instead? Van Fraassen seems to think so, in his earlier book at any rate, for he says that explanatory power is at least one of the factors guiding theory choice:

> When we decide to choose among a range of hypotheses, or between proffered theories, we evaluate each for how well it explains the evidence. I am not sure that this evaluation will always decide the matter, but it may be decisive, in which case we choose to accept that theory which is the best explanation. But, I add, the decision to accept is a decision to accept as empirically adequate.[40]

The advantage of this approach is that it does not quarrel with the practice of science; it does not tell scientists that they should not use Inference to the Best Explanation. Rather, it says that science should carry on as normal, but that we should be more judicious in our attitude towards its conclusions and preferred theories. The attitude should not be one of belief, but one of acceptance as empirically adequate. Taken in this way, constructive empiricism may seem eminently sensible; we should hedge our bets. Why stick one's neck out and believe in the truth of a theory when one can accept the same theories for the same reasons but take the less risky attitude of belief in empirical adequacy?

I shall argue in this section that this "have your cake and eat it" approach will not work. However, it is worth pointing out that anti-realists may choose to reject Inference to the Best Explanation altogether, even as a way of inferring empirical adequacy, and van Fraassen's later work suggests a move in this direction. But, in that case, the anti-realist has some hard work to do. It does seem that Inference to the Best Explanation does best describe the way in which scientists make inferences. So denying this will involve some tricky argument as to why the appearances are deceptive. But if the anti-realist accepts that Inference to the Best Explanation is the way scientists reason, then we need an explanation as to how it is that such a bad epistemology gives empirically adequate results (and I am about to argue there is no reason why Inference to the Best Explanation should give empirically adequate results, unless it gives ones that are close to being true). The only remaining position would be a more sceptical one, according to which we have no reason to think that science gives results that are even empirically adequate. In short, I think that constructive empiricism, rather than being a judicious midway point between unjustified faith and unproductive scepticism, falls between two stools.

As a quick way to see this, I will mention a criticism which van Fraassen levels at Inference to the Best Explanation. He asks, why should we think that the actual explanation of the evidence should be found among the theories we are considering, when there must be (infinitely) many theories that are also potential explanations of the evidence, but which we have not even thought of? The chances of our even thinking of a correct explanation must be small; most potential explanations are false. I will suggest a response to this criticism in a later section, but for the time being, it is worth noting how sceptical a criticism it is. Given any evidence there will be many theories that are consistent with the evidence and indeed (potentially) explain it, but which have consequences incompatible with observable, but as yet unobserved facts. For instance, the evidence of astronomers seems to be compatible with the hypothesis that the planet Venus has a thin mantle and the hypothesis that its mantle is thick. One of these is false and the difference is potentially observable (we could, in principle, go and find out). Furthermore, the false hypothesis is empirically inadequate, because it has false observable consequences. Generalizing, it is clear that of all the possible hypotheses which might explain or accommodate certain evidence, most will be not only false but will also be empirically inadequate. So, by van Fraassen's argument, we should think not only that it is unlikely that a *true* explanation is among the theories we have considered, but also that it is unlikely that an *empirically adequate* theory is among them. In which case we should neither believe nor accept our best theories. Indeed, we should be generally sceptical about them.[41]

I now want to show why it is that the constructive empiricist cannot piggy-back on Inference to the Best Explanation, choosing only to take a weaker attitude to its conclusions. Let T be a theory which states the existence of certain facts F which explain the evidence E. On the Inference to the Best Explanation model, if the existence of the facts F constitutes the best potential explanation of E, then we are (sometimes) entitled to infer the existence of F and so the truth of T. We take the best potential explanation to be the actual explanation of the evidence. In contrast, the constructive empiricist wants it that the most we may infer is the empirical adequacy of T, not its truth. But, if we allow that T may not be true and hence that the facts F may not exist, we allow that what we have may not be the explanation. For the facts F to be the explanation of E, F *must* exist – if continental drift is the explanation of mountain ranges and of fault lines, then continents must drift.

The explanation of a fact is some other fact or set of facts. And so Inference to the Best Explanation is inference to facts. Someone who employs Inference to the Best Explanation cannot but take a realist attitude to a theory which is preferred on these grounds. It should be noted that this is fully consistent with having a less than certain belief in T. Perhaps the evidence is less than overwhelming. One may have rather less than the fullest degree of confidence in T – there is room for doubt as to whether the facts F exist, perhaps something else is the explanation of E. One's attitude towards T is one of a degree of belief less than full, confident belief. A degree of belief in T is not the same thing as a belief that T achieves some lesser degree of correspondence to reality that truth, e.g. empirical adequacy. One may have a reasonably high degree of belief in T, yet also believe that if it is false then it is not even empirically adequate. Say we have several rival theories of the extinction of the dinosaurs – meteorite impact, volcanic eruption, and viral infection. Even if each appears to be a very good explanation, if one theory, say meteorite impact, is true and the others are false, we would not expect the false ones to be empirically adequate, since the observable aspects of a meteorite impact, of volcanic explosion, and of viral infection are all so very different (even if we are not in a position to make the relevant observations). The point of the remarks made in this paragraph is to resist an argument for anti-realism which, at the same time, suggests a reconciliation with Inference to the Best Explanation. The idea was that being a realist is risky. Even if we infer in a realist way, it still makes sense to reduce one's belief as regards the theory from belief in its truth to belief in its empirical adequacy (to the attitude van Fraassen calls "acceptance"). However, hedging one's bets in this way makes no sense if one's only good reason for believing a theory to be empirically adequate is that one believes it to be true.

Anti-realism and the structure of science

Constructive empiricists should have no truck with Inference to the Best Explanation. At the same time they cannot, I believe, make sense of scientific belief, whether realistically or anti-realistically construed, without it. Anti-realists tend to favour something like the hypothetico-deductive model as capturing the structure of inference. This makes sense – on the basis of this model anything confirming a realistically understood theory confirms it anti-realistically too. What I mean is this: let theory T be confirmed by evidence E, i.e. T (with observable conditions C) entails E. On the assumption, which we are granting the anti-realist, that all evidence is observational, the following will also be true: T is empirically adequate (with observable conditions C) entails E. So if

E is evidence for T it will also be evidence for the more limited claim that T is empirically adequate. (If what we see in cloud chambers is evidence for quantum mechanics because it is entailed by quantum mechanics, then what we see in cloud chambers is evidence for the empirical adequacy of quantum mechanics because it is entailed by that too.) Consequently, we should regard any evidence as supporting the empirical adequacy of a theory at least as well as its truth and typically better, since the truth claim is riskier (and, in addition, requires the existence of unobservable entities).

My own view is that we do have reason to believe in the existence of even highly unobservable entities and that the constructive empiricists' argument does not work because they have too limited a notion of evidence. In particular, we shall see that the very structure of science is evidence in favour of its truth. The inadequacies of the hypothetico-deductive model of confirmation have already been subject to our scrutiny, and we will not dwell on them all again. But there was one omission from that discussion which deserves emphasis. According to the omitted factor, the confirmation given to a hypothesis by a piece of evidence is a matter solely of the logical relations between the hypothesis, observable conditions, and that piece of evidence. No mention is made of any possible relationship among different pieces of evidence. But this is implausible. Say I am measuring the boiling point of a new synthetic material. It seems surprisingly low. This could be an interesting discovery about this new species of compound, so I entertain this as a new hypothesis. But it could also be owing to contaminated equipment. I could do a number of things – I could analyze the sample for the presence of contaminants or I could repeat the experiment with new carefully cleaned equipment. Let it be that the analysis showed no contaminants and the repeat experiment gave the same result as the first. Both these results change the confirmation given to the hypothesis by the first test. The test itself has not changed, nor has its logical relation to the hypothesis. Yet it now provides stronger confirmation than it did initially. The analysis ruled out an alternative hypothesis (contamination), thus raising the support given to this hypothesis. The repeat test also rules out the contamination alternative, in a weaker way, by making it redundant. It also helps by suggesting that the first result did not come about by some other one-off error, such as misreading the thermometer. What this shows is that whether a fact is evidence for a hypothesis and if so to what degree it confirms it is dependent on its relationship with other things we know.

The hypothetico-deductive model ignores the fact that structural features of hypotheses (and evidence) are evidentially relevant. Perhaps the anti-realist can adopt a different epistemology which does take these things into account. But it is unclear why an anti-realist should regard these things as significant. For it is difficult to see how anyone could regard structural properties as epistemically significant unless it is because they think these properties are explanatory virtues. It is a central feature of Inference to the Best Explanation that it emphasizes the holistic aspect of evidence. On the larger scale we saw that criteria involved in detecting or inferring a law should include not only that it does entail the evidence, but also that it should be integrated into a system of laws that displays both strength and simplicity. The lesson of this for constructive empiricism is that it is not true to say that empirically equivalent theories will also be evidentially equivalent. Two theories may have precisely the same observational consequences yet one be better supported by the observational evidence than the other, because that theory would be better integrated with our other best theories. That is, as a whole we have reason to think that these theories, being integrated into a strong but simple system, state something like the laws that there are. The evidence is not just the agreement of observational consequences with what is observed, but also the internal features of the overall system of putative laws. This is what we learned from the discussion of laws in Chapter 1 and of Inference to the Best Explanation in Chapter 2.

What would we expect science to look like if constructive empiricism were right? If it were right, what constrains our choice of theory should be evidence of empirical adequacy, i.e. observational success (although simplicity, and so on, may be virtues that make for ease of handling). First, consider that the world is full of diverse and apparently unrelated phenomena. Think for instance of the range of so-called electrical phenomena: electrostatics, lightning, electrolysis and electrochemistry, electric current, the photoelectric effect, the behaviour of cathode ray tubes, and so on. Observationally these look very different. The ability of a rubbed balloon to stick to the ceiling looks nothing like the production of bubbles in a salt solution attached to a voltaic pile. And neither of these looks much like lightning or anything like the photoelectric effect. Going on appearances alone no-one would think of linking these phenomena to the strange behaviour of magnetic lodestone or to beams of light. Since these are all observationally so dissimilar and unrelated, we would expect, if empirical adequacy is our aim, that we would construct one theory for each of these different phenomena. We would have, we hope, an empirically adequate theory of electrostatics that allows us to predict the observational behaviour of things when rubbed. We would also have a set of predictively accurate generalizations dealing with the behaviour of liquids into which wires attached to a pile are dipped. We might have a theory about the conditions under which lightning is to be observed. And so on; each set of phenomena would have its own empirically adequate theory.

Science, of course, does not look like that. All the phenomena in question can be subsumed under the mutually integrated theories of electromagnetism and atomic structure. A remarkable fact about modern science is that as the number of phenomena which science has investigated has grown, the number of theories needed to explain them has decreased. And those theories have been deeper and more general, and, correspondingly, more integrated when it comes to explaining the phenomena. Our theories are not low-level generalizations each specific to particular phenomena, but are broad laws that operate together to explain a wide range of phenomena. From the point of view of the constructive empiricist, who thinks our theories are empirically adequate to the phenomena but are most probably false, this fact should not be merely remarkable but really quite extraordinary. If those theories are false it should be a surprise that they should neatly integrate into a powerful but simple system. On the other hand, this fact is rather less extraordinary if those theories are close to the truth and are reasonable representations of the laws that there are, because integrating in this way is precisely what we expect of laws.

I have argued that the ability of a theory to integrate with other theories and its ability to produce novel and unexpected true predictions constitute evidence for its truth that goes beyond observational success. A response to both points is that such achievements in a theory reflect the criteria by which scientists select theories. It is no surprise that theories tend to integrate. A highly unified science is easier for the scientist to use, not to mention more aesthetically pleasing. From among equally empirically adequate theories, we would expect the scientist to choose the one which most displays this structure. While the key thing in theory choice will be the external or representational feature of empirical adequacy, internal pragmatic or aesthetic considerations may also play a part. This explains why our favoured theories possess desiderata such as simplicity and the ability to generate novel predictions. As successful predictions are just what constitutes empirical adequacy, and as empirical adequacy is the key feature in theory choice, it is scarcely surprising that theories should continue to possess it, for a time at least.

This selection argument can be countered with three rejoinders. The first is that the selection argument is disingenuous. Assume the realist were right and that unity, simplicity, strength, and so forth, are evidence for truth. Then it would make sense for scientists to select theories with these characteristics. But this would not be just to think of the selection argument as *undermining* the claim that these characteristics are a sign of truth.

The second rejoinder accepts that, if creating theories with unity and with successful novel predictions were easy, the selection argument would have some force. But selection for unity is not easy. By looking at particular cases we find the constructive empiricist position less plausible. The fact that scientists may prefer unity and integration does not explain why it is possible to find satisfactory theories with such properties. It is not easy to integrate our theories, as Einstein and those seeking to unify relativity and quantum mechanics have found. When we do so it is something of an achievement. And, for reasons I have made clear, it is regarded as a sign that science has got things by and large right. If the world, the underlying unobservable world, is not integrated, or is integrated in a way quite unlike that in which we believe it to be, then we would not expect to be able to integrate our theories as we have been able to do. The fact that we can is a sign that the world is integrated this way, and so is a sign that our theories are true, or nearly so.

That scientists themselves take unification to be a sign of truth is clear. Einstein has this to say about one step on the road to unification:

> Since Oersted discovered that magnetic effects are produced not only by permanent magnets but also by electrical currents, there may have been two seemingly independent mechanisms for the generation of magnetic field. This state of affairs itself brought the need to fuse together two essentially different field-producing causes into a single one – to search for a single cause for the production of the magnetic field. In this way, shortly after Oersted's discovery, Ampère was led to his famous hypothesis of molecular currents which established magnetic phenomena as arising from charged molecular currents.[42]

Ampère's hypothesis is a good example of a hypothesis adopted and given credibility independently of its possessing empirical support, and so independently of its demonstrated empirical adequacy. Ampère conjectured that the field of a magnet is created by the rotation of charges within the matter of the magnet. Experimental verification of the hypothesis was very difficult – it was 92 years until de Haas and Einstein were able to demonstrate the Richardson–Einstein–de Haas effect that the hypothesis predicts. As Ampère himself wrote, many years before:

> Those periods of history when phenomena previously thought to be due to totally diverse causes have been reduced to a single principle, were almost always accompanied by the discovery of many new facts, because a new approach in the conception of causes suggests a multitude of new experiments to try and explanations to verify.

That the facts are not easy to produce shows that they are not trivial. Their production is powerful evidence that the theories have latched onto the truth, that the entities they postulate exist, and that they provide genuine explanations. Returning to Dalton's atomic hypothesis with which we started this chapter, the phenomenon that eventually clinched the debate in its favour was the ability of the theory to account for the phenomenon of isomerism, which was discovered some considerable time after Dalton propounded his view. (Isomers are compounds sharing the same formula but having different chemical or physical properties; this is easily explained in terms of molecules having different arrangements of the same set of atoms.) The fact that a new theory can capture previously diverse phenomena, can unite independent older theories, and can predict otherwise unexpected new phenomena simply cannot be explained adequately by the selection argument.

The third rejoinder is an extension of the first and second. It argues that the status of evidence is not impugned by the procedures by which it was obtained. Imagine a detective who is not interested in discovering the truth but only in proving the case against the prime suspect. In his zeal he discovers and presents to the court some very persuasive DNA evidence. Then the defence demonstrates that the detective is motivated only by the desire to secure a conviction. This may cast doubt on the evidence, since it might be thought that the detective fabricated it. But if the prosecution can show that the detective could not have fabricated the evidence, then the evidence may still stand. This would be the case even if the court knew that the only reason that evidence had been discovered was the determination of the detective to find it.

The point of this example is that if unity, strength, and simplicity are evidence for the truth of a theory, then their being evidence is in no way undermined by the fact that they are selected for, whatever the motivation (whether it be for love of truth, or convenience, or beauty) behind the selection. The way in which evidence confirms a theory is quite independent of the ways in which the theory and the evidence were obtained (it would make no difference if the theory were pulled out of a hat). In particular, I have argued that it is part of our concepts of law, explanation, and confirmation that simplicity, strength, and the ability to integrate with other theories are properties which, together with observational success, count as evidence for a theory.

Criticisms of Inference to the Best Explanation

Anti-realism is motivated by epistemic concerns. Theories appear to postulate the existence of things that we cannot observe. In the earlier parts of the chapter, when discussing instrumentalism, we saw that one response to this is to reconstrue the meaning of theories so that it is no longer the case that they imply the existence of unobservables. In this way the problem is supposed to be dissolved away. However, as we have seen, the reconstrual just does not work. And so later versions of anti-realism, like van Fraassen's constructive empiricism, take theories at face value. The response to the epistemic problem of unobservables is, in essence, to give in to it. We cannot know theories to be true and we should not believe them to be so; we should take a weaker attitude to theories, namely one of acceptance – believing them to be empirically adequate. This is clearly sceptical, but the trick is to show that this scepticism is limited; later in this section, I shall suggest that it is not. The argument between the realist and the anti-realist is now about the power of the epistemic apparatus of science. The realist thinks it is stronger than the anti-realist takes it to be. In particular, the realist thinks that Inference to the Best Explanation has the power to deliver knowledge of, or at least justified belief in, unobservables; the anti-realist denies this. Earlier on, I argued that the anti-realist cannot expect Inference to the Best Explanation to deliver grounds for acceptance if it does not also deliver grounds for belief. Some tools give you all or nothing. Now I want to turn to the anti-realist's arguments for thinking that Inference to the Best Explanation cannot deliver good grounds for belief.

Van Fraassen points out that Inference to the Best Explanation can only be any good if we have reason to think that we have thought of the best explanation. While the evidence could allow us to make a comparative judgement among those theories we have considered, it cannot tell us whether the truth is likely to be among them. Indeed, argues van Fraassen, it is likely not to be. For there must be many theories, perhaps not yet conceived of, which explain the evidence at least as well as our best theory. Most of these will be false, so it is reasonable to conclude that ours is likely to be false too. And so, van Fraassen concludes, we should reject the rule of Inference to the Best Explanation.

In fact I agree with van Fraassen that there is no good rule of Inference to the Best Explanation. Earlier on I emphasized that there is no model of explanation or confirmation. Correspondingly, there is no *rule* of Inference to the Best Explanation. I shall reiterate the point in the final chapter, where I argue that there is no such thing as the scientific method. But to say that Inference to the Best Explanation is not a rule is not to say that it is no good at all. For it is true that at an early stage in an investigation we might reasonably think that we have not thought of all the plausible explanations. Where data are few, we might think that we have not got enough evidence to make a satisfactory choice between those explanations that we have thought of. For instance, Israeli scientists have recently added a new potential explanation, radiation from colliding neutron stars, to the old array of hypotheses about the cause of the extinction of the dinosaurs. And, while what evidence there is disposes many in favour of the meteorite hypothesis, the evidence is by no means conclusive. Hence it may be too early to believe that the best explanation we have is the true one.

Nonetheless, we may one day get to that point, as I believe we have done in other cases (the atomic hypothesis, the electromagnetic theory of light, and so on). Just because there are situations in which Inference to the Best Explanation cannot be used, does not mean there are no situations in which it may. How do we know when we have got to that point, when we think we have considered the full range of possible explanations and have enough evidence to decide between them? I will address this question again in the final chapter. But the short answer is that we cannot lay down rules for this. This is a judgement scientists have to make.

To make this judgement scientists will have to take into account a whole raft of factors. For a start the very nature of the question they are dealing with will determine the confidence the scientists have in believing that they have thought of all plausible explanations. The demise of the dinosaurs is remote in history. It is an event quite unlike any we have witnessed. So we are not sure, in advance, what the explanation might look like or whether we will be able to gather sufficient evidence to decide the issue. By contrast, a detective investigating a suspicious death will have a fair idea of what the plausible explanations might be and how to get the evidence that may discern between them. Then there is the nature of the evidence itself. How reliable is it? Are there any unexpected predictions explained by one theory but not the others? There will also be extraneous factors. How thoroughly has the issue been investigated? How eminent and successful are the proponents of the theories? Are they sufficiently expert in their fields? (The meteorite explanation of dinosaur extinction met with scepticism at first on account of the fact that its proponent, Luis Alvarez, is a physicist not a palaeontologist.) Sometimes scientists may not be able to say much more about why they are convinced that a certain theory is the right explanation, except that they feel that it must be. Even this is not irrelevant if a particular scientist has some history of success.

These considerations are all inductive. Scientists (both as individuals and as a community) have learned through experience the conditions under which it is sensible to use Inference to the Best Explanation. Since, as I have been claiming, Inference to the Best Explanation is itself the principal vehicle for inductive inference in science, it would follow that the application of one inductive inference (Inference to the Best Explanation) depends on the application of another. This seems to threaten a regress. Another way to see this is to consider how we might rationally justify our practices regarding the application of Inference to the Best Explanation. We might say that we have acquired these practices because we have found them to be the most successful, and the best explanation of this is that they somehow match the structure of the world, they are truth-tropic. So the employment of Inference to the Best Explanation would depend on an application of Inference to the Best Explanation. And that application would itself depend on a further prior application of Inference to the Best Explanation. Again a regress threatens.

This regress is similar to that generated by Goodman's problem. There we had to choose between inferences using grue and those using green. The answer is that experience has taught us that green is a natural kind and that grue is not. But this discovery about natural kinds itself must be the product of an inductive argument. So, deciding which inductive inference to make depends on a previous inductive inference. In which case, where do we start? We tackled this problem in the last section of Chapter 3, but I shall recap briefly here. There are two parts to the answer. One is that we start with inductive habits and dispositions that in themselves do not depend on any conscious previous induction. These habits may be innate but, in the case we have been looking at, they may be learned in the process of learning to become a scientist. The fact that we may start from a habit for which we do not, or even cannot, give an explicit justification does not mean that it cannot provide a basis for knowledge. I will argue for this claim in Chapter 7. Secondly, the habits in question, the acquired ones at least, are self-correcting and improving, and ultimately self-justifying. This is because we can check our habits to see that they are working, i.e. yielding the results we want. If not, we learn to place less reliance on the habit and seek adjustments that may yield greater reliability. If we do get good results our use of the habit is reinforced. In some cases, such as our perceptual habits and some of our inferential habits, we may investigate their structure and decide by using science (e.g. physiology) whether they are reliable. Either way we will be using inductive practices to calibrate, investigate, and even justify our inductive practices. Certainly this process is circular, but whether the circularity is vicious (as a Humean sceptic will claim) or not is another question, which also will have to wait until Chapter 7.

The preceding remarks may also help us tackle some other of the objections which may be raised against Inference to the Best Explanation. Two objections are of particular interest. Peter Lipton calls them "Hungerford's objection" and "Voltaire's objection." Hungerford's objection is that beauty is in the eye of the beholder – the goodness of possible explanations is subjective, and so the choice of the best explanation also be subjective, and for that reason not a suitable guide to truth.[43] Voltaire's objection is that this may well not be the best of all possible worlds, i.e. why think that the best explanation is true? Might not the world be constructed in such a way that bad (i.e. complicated, unsystematic, weak) explanations are true?

A few paragraphs back I said that the answer to the question "How do we know when to apply Inference to the Best Explanation?" is *experience*. Experience is the answer to these objections too. Indeed, both are related to that question. Take one desideratum of good explanations, simplicity. Simplicity is often criticized as being too subjective a notion to play a part in scientific inference (or the concept of a law of nature). I think this is a mistake. For a start, it is clear that simplicity is one of the factors a scientist seeks in an explanation. And so it had better be that it is not merely subjective, lest we end up with another kind of scepticism about objective scientific knowledge. People may have differing views as to what is simple, but that does not show that simplicity is subjective (any more than the fact that people disagree about what is true shows that truth is subjective). Rather it may be that simplicity is context dependent, or that what counts as simple for the purposes of explanation and inference is a matter of discovery. I suspect that both these are true. Take the thought that discovery tells us what is simple. Say we accept that it is experience which tells us that when it comes to emeralds we should make our inferences using "green" not "grue"; it seems to me that it is just this experience which backs our intuition that "all emeralds are green" is a genuinely simpler hypothesis than "all emeralds are grue".

Turning now to Voltaire's objection, it seems that the best reason for thinking that good (potential) explanations will turn out to be actual explanations, i.e. veridical, is that experience tells us so. Earlier I dismissed the success-of-science argument for realism (the truth of theories) as just an amalgamation of the individual arguments for each particular theory. Nonetheless, if the success of science can be explained, then the best explanation of it is the reliability of our inductive practices. If it is right that the inferential practices of sciences are best described as Inference to the Best Explanation, and if one accepts that science has been very successful at producing theories that have been shown to be true, then there is a lot of inductive evidence that Inference to the Best Explanation is truth-tropic, i.e. that it tends to yield inferences to true conclusions. There are four rejoinders to this justification of Inference to the Best Explanation. (a) The inferential practices of science do not conform to Inference to the Best Explanation. (b) The evidence is not as positive as it seems – the application of Inference to the Best Explanation in the history of science has been one of failure just as much as success. (c) The success of science is the observed success of science, so the success of science can only justify Inference to the Best Explanation insofar as the inferences concerned have observational conclusions; there is no reason to think that inferences to the existence of unobservable entities is justified. In addition, (d) this justification of Inference to the Best Explanation is blatantly circular – it seeks to justify an inductive practice using induction.

To rejoinder (a) I may best suggest that the reader consider whether scientists, and indeed the rest of us, use Inference to the Best Explanation. Rejoinder (b) is more difficult, as it seems to require a survey and assessment of the success of theories throughout the history of science (modern science at least). To some extent I shall try and address this question in the final chapter of the book; for the time being, the following general consideration may suggest that science has been more successful than not. Science is by and large cumulative, it builds on what has gone before (although admittedly it throws out much as well); in addition, it is progressive, at least in the weak sense that scientists do not deal with exactly the same old phenomena as they always have. Rather, they are for the most part investigating new phenomena, entities, even whole new branches of science. Accumulativeness and progress are connected in that it is by building on older accepted theories that new phenomena can be investigated. It is by accepting Dalton's atomic hypothesis that we can go on to ask questions about the structure of atoms, and by accepting Mendel's genetic theory that we can think about identifying genes for particular hereditary traits. These remarks are supposed to be weak descriptions, in that I do not here assume that accumulation is the accumulation of truths or knowledge, or that progress is improvement. Accumulation is just adding to what has gone before and progress is just the development of new areas of inquiry. Nonetheless, it seems implausible that there could be accumulation or progress in the weak sense without at least a fair degree of truth or proximity to the truth. Were the foundations and materials of accumulation and progress rotten to the core we would expect the structure built upon them to collapse. It would be odd to think that Mendel's genetics were wildly mistaken, yet it be possible to do something that scientists (mistakenly therefore) think of as "looking for the gene for sickle-cell anaemia" (let alone something they think of as "discovering the gene"). Surely such sophisticated investigations would have foundered long ago were their fundamental assumptions awry.

Rejoinder (c) says that we have not observed the success of instances of Inference to the Best Explanation which have the existence of unobservable entities as their conclusions, since *ex hypothesi* any success or failure of such inferences is unobservable. So the success of science could at most justify the use of Inference to the Best Explanation to infer the existence and nature of things which, although unobserved, are potentially observable. This would be a powerful argument if that which was once unobservable remained thus evermore. But that is not the case. Inferences to the existence of unobservables have later been verified by direct observation once observational techniques have improved. We can now observe microbes and molecules, the existence of which was once a purely theoretical, explanatory hypothesis.

Rejoinder (d) is of course just Hume's problem of induction. Insofar as Inference to the Best Explanation suffers from Hume's problem, Inference to the Best Explanation is not worse than any other form of non-deductive inference. So as long as an anti-realist is not also a sceptic, his or her epistemology will suffer from this problem also. Say we were to ignore Hume's problem for a moment. Clearly using Inference to the Best Explanation to justify Inference to the Best Explanation is not going to convince an anti-realist who has serious doubts about Inference to the Best Explanation to begin with. But if someone already does accept Inference to the Best Explanation then they may use it to defend Inference to the Best Explanation against objections like Voltaire's. For if they think that Inference to the Best Explanation is good for justifying empirical beliefs in general, they should also think that Inference to the Best Explanation would be good for justifying a particular empirical belief, namely that explanatory power is a guide to truth. As I have already promised, in Chapter 7 I will explain why we are entitled to ignore Hume's problem and why using Inference to the Best Explanation to get to know about the applicability and reliability of Inference to the Best Explanation is not viciously circular.

Further reading

Bas van Fraassen's constructive empiricism is clearly and forcefully advocated in his *The scientific image*, while Nancy Cartwright's anti-realism about laws is to be found in *How the laws of physics lie* (especially the Introduction and Essays 2 and 3). Larry Laudan's "A confutation of convergent realism" contains some strong arguments against the view that science is convening on the truth. Inference to the Best Explanation is defended in Peter Lipton's *Inference to the best explanation* (Chs. 7 and 9). The literature on scientific realism and its critics is enormous. A useful pointer is Richard Boyd's "On the current status of scientific realism". Influential writings include Arthur Fine's "The natural ontological attitude" and Ian Hacking's stimulating *Representing and intervening*. Many of the papers mentioned are collected in David Papineau's *The philosophy of science*.

Part II

Reason

Chapter 5

Inductive scepticism

In Part I, I argued that we could have reason to believe that science, or at least reasonably large chunks of it, does provide a fairly accurate representation of aspects of the world. More precisely, we are justified in believing that our best theories have a high degree of what is called *verisimilitude* in their representations of the laws of nature and other law-like facts. Two theories can both be strictly false, yet one can be better than the other by virtue of being closer to the truth – which is expressed by saying that it has greater verisimilitude. We will need to take account of this.

However, there is a more immediate obstacle that needs clearing before we consider those rather more subtle issues. This is the rather large obstacle of the problem of induction. How could I reach the positive conclusion I did about the verisimilitude of theories without considering the objection raised by Hume to the possibility of gaining any general or universal knowledge from evidence consisting only of particular things and facts? The short answer is that my argument was, in outline, that if we had reason to think that our theories are empirically adequate we would also have reason to think that our theories are true or nearly so. This invites the question, for both the constructive empiricist and the realist, of how we can have knowledge of a general sort at all, whether that knowledge be of empirical adequacy or of verisimilitude. So, either way, the problem of induction needs to be addressed, and that is what I will do here.

Hume's problem revisited

Hume's problem, you will recall from the Introduction, is that there seems to be no non-circular justification of our claim to general knowledge or particular knowledge of events that we have not observed. I will restate the problem, in slightly different forms, to remind us what we are dealing with. The general problem is this: given an argument from premises about what is observed to a conclusion about what is not, how can those premises ever justify that conclusion?

(Hume believed that premises about the observed never entail anything about the unobserved, and so the justification cannot be deductive. As we shall see later on, our observations are typically *theory laden*. As an observation may imply some theory it will imply something about the unobserved. A simple example is this. I may report what I observe by saying that there is a lemon on the table in front of me. If this is true, then certain things follow, e.g. (a) that there is some flesh behind the skin, and (b) that the object in front of me originated from a lemon tree. But statements (a) and (b) are statements about things I have not observed. For this reason some positivists (such as those promoting the instrumentalism discussed in Chapter 4) have wanted to deny that "I see a lemon" could be an observation report. They maintained that even this statement must be inferred from some prior observation statement such as "I see an object with a yellow surface, etc", or even "I am having an experience of seeing an object with a yellow surface, etc.". Be that as it may, even if we accept that *some* observation statements entail propositions about the unobserved, it is certainly the case that the inferences employed in science are such that their premises do not entail their conclusions. The claim that dinosaurs were made extinct by a meteorite is not entailed by the evidence gathered by the proponents of that theory, nor do our observations entail the proposition that Halley's comet will next return in the year 2052, which astronomers predict with confidence.)

So to return to our question (slightly modified): How can premises about what is observed justify a conclusion about what is unobserved, when the former do not entail the latter? (In what follows I will take "unobserved" in a broad sense, to include what is present but distant or hidden, what is future, what is imperceptible, and what is general rather than particular, including facts about universals.) Consider two possible justifications of arguments from the observed to the unobserved:

> Justification A: Some such arguments justify their conclusions since our experience shows that arguments of that form tend to lead to true conclusions from true premises.
>
> Justification B: Some such arguments justify their conclusions since their premises, along with the additional uniformity premise (that the unobserved resembles the observed) entail their conclusions.

Justification A proceeds by arguing that the form of inductive arguments is reliable. Justification B operates by adding a premise that turns the inductive argument into a deductive one. Neither justification succeeds. Justification A tells us that the past success of inductive reasoning is a reason to accept it in this case – but that is just a case of inductive reasoning, which is what we are seeking to justify. Justification B assumes the uniformity premise. Why should we think the uniformity premise is true? Experience shows that the world is uniform. But this experience justifies accepting the uniformity principle only if we make use of the uniformity premise. Both cases suffer from circularity.

Hume seems to have thought that the only conceivable justification for an inductive argument would be something like justifications A and B. Consider B. Is there any substitute for the uniformity premise, i.e. an alternative additional premise that would turn the inductive argument into a deductive one?

What such an additional premise might look like will depend on the case, but there will be certain features that such premises will typically have in common. Note that the existing premises are statements about what has been observed. And the conclusion is a statement about the unobserved. So the additional premise must *mention the unobserved*; because premises that make no mention of the unobserved cannot entail a conclusion about the unobserved. No amount of information about what I did today entails any fact about what I will do tomorrow. To make an entailment of that sort I need to add a premise that mentions both today and tomorrow – for instance, that I will spend tomorrow morning as I spent this morning.[44]

The additional premise will also typically be known only *a posteriori*. Interesting statements about the unobserved are not ones I can know *a priori* – pure reason alone cannot tell me anything about tomorrow's weather. I make the provisos "typically" and "interesting" as some propositions about the unobserved can be known *a priori*: that I am not observing the unobserved, for instance. But such *a priori* propositions are not going to get us far in science. So, although I know *a priori* that it will either snow or not snow tomorrow, that will not tell me whether any empirical proposition is true (e.g. whether the lawn will look green or white).[45]

So the additional premise is a proposition about (in part) the unobserved, which to be known can be known only *a posteriori* (what I shall call an *empirical* proposition about the unobserved). Note that the uniformity premise is just such a proposition. It mentions the unobserved and it cannot be known *a priori*. But empirical propositions about the unobserved are precisely the sort of propositions that one needs induction to know or to have a justified belief in.

Let us illustrate the previous point. I may tell you that I expect this piece of sodium to burn with a yellow flame. You ask me why I expect this. Because, I say, I have previously seen sodium burn with a yellow flame. You may then ask, in a Humean spirit, why do my previous experiences of a yellow flame justify my expectation in this instance? I might appeal to the uniformity premise, which together with my past experiences entail the expected outcome. But I might not. I might make use of a more limited claim. I might appeal to the fact that the flame colours of metals are a stable and characteristic feature of them. This fact does the work just as well as the uniformity premise. But like the uniformity premise it is an empirical proposition, which is in part about the unobserved. So it stands in need of justification. There are two things I might say. Either I might point to my experience of the fact that copper always combusts with a characteristic green flame, that potassium burns with a lilac flame, calcium with a red flame ..., etc. This justification is clearly inductive. Alternatively, if I know enough about chemistry I might say the following: "Combustion involves the excitation of certain electrons in the outer shell of the atoms. When the electrons return to their ground state they emit the energy of excitation in the form of light. The ground and excitation states are fixed for each element, but differ between elements. Hence the wavelengths of light emitted are both fixed for each element and characteristic of it." The details do not matter. What does matter is that this new justification consists of further empirical propositions about the unobserved – this time even more general than the one they are justifying. Like that, they too stand in need of justification. So, even if we avoid circularity, we are faced with an infinite regress of ever more general claims about the unobserved.

One point should be mentioned: Hume assumes that empirical propositions about the unobserved stand in need of justification if they are to be credible, and their justification is a matter of inference from other propositions. This is one of the assumptions characteristic of *empiricism*. Some philosophers may wish to challenge this assumption. Be that as it may, I think it holds for the cases we are interested in. The sorts of regularity and law referred to in the previous paragraph are not the sorts of proposition one could rationally believe without justification stemming, as in those cases, either from the evidence of particular observed cases, or from other laws that themselves require justification. No one could think that the laws of quantum chemistry could be known by intuition or by virtue of their self-evidence. Nonetheless, a philosopher might suggest that we have to assume without justification some very broad principle, such as the uniformity premise. He or she may argue that we have to take the uniformity of nature as the unjustified assumption of all our inductive reasoning – a sort of grand article of scientific faith. There are two reasons for rejecting this. The philosophical reason is that this is just a cop-out. It would signal defeat by the sceptic while lacking the grace to surrender properly. It would also corroborate the assertion of the creationists that mainstream science no less than creationism rests on scientifically unjustifiable basic premises.

The non-philosophical reason is that in any case the uniformity premise is manifestly false. It is simply not true that the unobserved always resembles the observed. In particular, it is almost never true that what is future is just the same as what is past. Tomorrow is most unlikely to be exactly like today, and next month is certain to be radically different from last month. Tomorrow will be either colder or warmer than today, perhaps not a lot, but it will not be identical. Tomorrow I shall wear different clothes and the Moon will be in a different location. More leaves will have fallen from the trees and atoms that have been stable for millennia will have decayed. Many people will be dead who were for a long time alive, and others will come into existence. It is foolish to expect the future to resemble the past in a whole host of regards, and this fact is well known to us all, although it may be better known to good poker players than it is to most philosophers. The only way the future can perfectly resemble the past is for nothing to change. My point is not to heap further scorn on induction – far be it from me to do that. Rather, it is that the uniformity premise is not the thing to save it. Of course, we could make the uniformity premise true, restating it as "The unobserved resembles the observed in certain respects, but not necessarily all". That indeed is true. But it is now no good for its intended purpose. The restated uniformity premise no longer combines with the premise "Sodium has hitherto been seen to burn with a yellow flame" to entail the desired conclusion "This piece of sodium will burn with a yellow flame". For that we need yet a further premise to the effect that the colour of a metal's flame is one of the "certain respects" (unlike the colour of leaves in autumn) which the restated uniformity premise asserts to remain constant across the observed and unobserved.

For these reasons I think it is both unwarranted and erroneous to adopt the uniformity premise as an article of faith. For the second of the two reasons, its inability to be both true and useful, it seems pointless to try and justify it or anything equivalent to it. My own view, which I shall amplify later, is that it follows that there can and should be no generalized justifications of induction.

This leads to the following question: Is there anything

interesting that one might usefully say about the general form of inductive arguments? For one might think that, if there were, then one ought to be able to provide a general justification for them. This question arises for other reasons. First, we have talked about Hume's problem in terms of inferring from the observed to unobserved on the basis of resemblance, which in the Introduction I called "Humean induction". What about inference to the best explanation? If Hume's problem does not affect the latter, then we may have a non-Humean form of induction immune from Humean scepticism. Furthermore, in Chapter 1 I suggested that the problem of Humean induction is insoluble unless one takes laws to be relations of universals, in which case Humean induction can itself be interpreted as a case of inference to the best explanation. So, optimistically, we may hope that even Humean induction may be saved, albeit indirectly. Sadly, this is not the case. I argued in Chapter 1 that inference to the best explanation also falls victim to scepticism. Hume's problem does not depend on the inductive inference being made in terms of resemblance. So long as the inference is from the observed to the unobserved, it does not matter how the inference was carried out, the inference cannot be justified in the way that deductive arguments are justified. Any such justification will imply an empirical and general relation between the observed and unobserved, one which will itself be in need of justification just as much as the inference it sought to justify. Say I use an inference to the best explanation, as for instance Dalton did in inferring the existence of atoms. As Dalton found, it is a contentious inference. What reason have I for thinking that it will lead me from true premises, my experimental evidence, to a true conclusion about unobservable atoms? Whatever my reason is, it must be a reason for thinking that an argument like this will tend to lead from true premises to a true conclusion. For instance it might be that inferences to the best explanation, of this kind at least, have hitherto been successful. I might argue either on the basis of resemblance of the current argument to its successful predecessors or on the general grounds that the best explanation of the success of inference to the best explanation is that it has a tendency toward truth. Either way I will commit myself to a proposition as replete with reference to the unobserved as the proposition I set out to justify. Hume's argument is perfectly general with regard to both the sort of inference we may wish to justify and the sort of justification we may try to employ to do the justifying, because whatever could do the job of justifying an inference from the observed to the unobserved must itself be an inference from the observed to the unobserved.

Clearly Hume's problem needs to be addressed by a philosopher of science. Are not, as I suggested in the introduction, the empirical sciences distinguished from, for example, arithmetic, geometry, logic, metaphysics, and theology, by the fact that they are *inductive* sciences? If Hume is right about induction then there is no scientific knowledge. Scientists are never justified in believing a theory. Science is not the rational enterprise we supposed it to be. In the face of such conclusions there are several possible responses. One is to embrace the conclusions boldly, and some philosophers have even done so. Paul Feyerabend, who was fond of putting nuclear physics and voodoo on a par, is one who did so. One can embrace the conclusions in part. Sir Karl Popper argued that, although we can never be justified in believing a theory, science can still be rational. We shall see that this course between Hume's problem and anti-rationalism cannot be steered successfully. Given the apparent failure of rationalism about science (the view that science is rational), the following question is raised: What, if it is not reason, is it that makes scientists do what they do? Pastures new are opened for historico-sociological explanations of scientific practice. This is interesting with regard to creationism. For, if we reject the idea that science is able to latch on to objective truth, then we will have to give up the privileged status of mainstream science relative to creation science. The feeling that creation scientists are just operating in a different paradigm to the mainstream may account for their being regarded with less than full derision in some circles.

On the other hand, one might not be so willing to give up on inductive science. In which case one will have to find a flaw with Hume's reasoning. One way to do this is simply to show him to be wrong by doing what he says cannot be done – giving a non-inductive justification for induction. We shall look at two cases of this. The first is Sir Peter Strawson's argument that there are grounds which can be cited *a priori* as justifying an inductive conclusion. The second is an instance of the claim that inductive arguments carry with them *probabilistic* justification. Neither of these claims that inductive arguments can be deductively valid. Nor do they clearly or openly depend on inductive reasoning. Nonetheless, we shall see that both are inadequate to Hume's challenge. Hume's argument, it seems, is right and there is no joy to be had in this direction. Nonetheless, I do not think the case for science is hopeless, and in Chapter 7 I hope to show why.

Theory and observation

Before looking at the various approaches to coping with Hume's problem I want to raise two further sceptical questions surrounding science and induction. They both concern the relation between theory and observation, and often go together under the title *theory ladenness of observation* (or *theory dependence of observation*). Nonetheless, the two questions are importantly distinct. The first deals with the meaning of observation statements. The second deals with the question of when such statements are true. The first I shall call *meaning dependence of observation on theory* and the second *cognitive dependence of observation on theory*. Meaning dependence of observation on theory notes that the meaning of a theoretical term depends on the theory in which it plays a role. This is plausible. To explain to someone what "phlogiston" and "gene" mean I would need to explain to them the phlogiston theory of combustion or something about genetics. As the gene example shows, a term can be both observational and theoretical at the same time. The problem comes when we consider that this is true of all observational terms. This might be argued for in two ways. The way that this claim is usually defended is to take a term that seems as untheoretically observational as one could expect and show that understanding it involves grasping a theory, albeit a low-level theory. "The leaf is green" is an observation statement of the most basic sort. How should we explain "green" to someone? One might use a sample: "To look like this is to be green" (said pointing at a green object). But this is not sufficient for a full understanding. For not everything which looks that way is green (e.g. a white object under green light), nor does every green object look green (e.g. a green object in the dark). Furthermore, one is disinclined to say that the chameleon *is* green or that the heat-sensitive image *is* green because their appearance is not constant. For these reasons philosophers (some at any rate) have suggested that calling something "green" is to attribute to it a property that causes an object to appear green (or causes us to sense it as green) when viewed under normal conditions. If this is right, the term "green" is partly theoretical – it involves the theory that objects can possess certain properties which have specified causal roles.

We briefly encountered cognitive dependence of observation on theory earlier in this chapter. It poses a different problem. Here the claim is that knowing any observation statement to be true depends upon knowing some theory. That something like this is true for much of science can be seen by considering the fact that most observations in science are made using sophisticated instruments, which have been designed using a great deal of theory. Even the design of an everyday ammeter presupposes a good deal of electrical theory. At the other end of the scale I mentioned the observation of neutrinos in a vast bath of detergent. The most esoteric theories of subatomic physics went into the design of this observational apparatus and the interpretation of its output. But even at the least sophisticated end of the observational spectrum some theory is assumed in making an observation, even if it is just the assumption that conditions for observing are normal and propitious for making an observation. This thought should be reinforced by reflecting that even unaided visual observation in science is not easy. For instance, merely reading a thermometer may easily be upset by parallax errors or distorting refraction caused by heat.

So, according to cognitive dependence of observation on theory, our knowing any observation statement to be true depends on the truth of some theory. This reverses the normal cognitive dependency, which is that our knowledge of the truth theories depends on observation. After all, that is what observation in science is for. So we cannot know the truth of theories without observation, and we cannot know the truth of observations without theories. Where, then, do we start?

Cognitive dependence of observation on theory and meaning dependence of observation on theory need to be distinguished from one another (though not all philosophers have succeeded in doing this). An observation may be cognitively dependent on one theory and meaning dependent on another. For instance, if we observe a neutron star with a radiotelescope, that observation is cognitively dependent on the theory of radiotelescopy. If we state our observation using the phrase "neutron star", then that report is meaning dependent on the theory of stellar evolution.

The Duhem–Quine thesis

Cognitive dependence of observation on theory says that we cannot make an observation independently of some theory. As we use observations to test theories, this means that we cannot test theory A without depending on some theory or auxiliary hypothesis B. Say we set out to test theory A. In drawing our conclusions about what we will expect to observe, we must make use of B. The expected observation is a consequence of A and B together. What happens if we in fact observe something which conflicts with our prediction? Is this decisive evidence against A, the theory we are testing? It need not be decisive, because our prediction was derived from A and B together. The falsity of the prediction may be explained by the falsity of B rather than the falsity of A. All we know for sure is that A and B are not both true together. This means that theories do not get tested singly, but together. Perhaps we may try and test B independently of A. But cognitive dependence of observation on theory says that we will have to employ some other theory, C, in the testing of B. So, in effect, all these theories are being tested together. In Quine's words "our statements about the external world face the tribunal of sense-experience not individually but only as a corporate body".[46] One consequence of this, which Quine emphasizes, is that any hypothesis apparently falsified by observation can be retained as true, so long as we are willing to make appropriate changes elsewhere in the system of beliefs and theories.

This might appear to lead to a new sort of scepticism. Helen is testing theory A, in conjunction with theory B, while Richard is doing the same. They both find a prediction falsified. So Helen makes a change to theory A, while Richard adjusts theory B. Is there anything that would decide which of the two is right? And, if nothing can choose between them, we will now have two competing theoretical systems which are judged equally well at the tribunal of experience. A proliferation of such equally successful systems seems inevitable.

Quine and others have suggested that the rule is that we should make those adjustments that cause least disruption to the existing system of beliefs. The question remains though, whether we can justify such a principle of choice, or whether it is simply pragmatic – choosing the path of least resistance for maximum convenience.

One answer to this problem is furnished by inference to the best explanation. Quine's holistic view, that our theoretical system faces the evidence as a whole, clearly mirrors my remarks in Chapters 1 and 2 about laws and explanation. Adapting the Ramsey–Lewis view of laws within a criterial approach, I argued that our grounds for taking something to be a law or an explanation would depend on our fitting the proposed law or explanation into what would be the overall system that optimally displayed strength and simplicity. This I used to explain the desiderata of inference to the best explanation. We can use this approach to explain why, typically, we choose to make those changes which least disrupt the overall system of beliefs. This is because they will be the changes that lead to the simplest revised system. Put another way, the theory that undergoes least revision will be a better explanation than the one that is badly disrupted by changes.

It may seem that, in the context of inference to the best explanation, the preference for least disruption is obvious. This need not be the case. It may be that a system is adjusted by a series of steps, each of which is the least disruptive step to take at that point; yet it may nonetheless be the case that a simpler system would have been obtained by making one drastic change somewhere along the line. We need to be open to this possibility, which will be discussed in more detail in Chapter 8.

Of course, inference to the best explanation is only a solution to these questions if we know that we may rely upon it. But, until there is something like a solution to Hume's problem, we do not know that we are entitled to take this approach.

Falsificationism

Before considering attempts to tackle Hume's problem head on, we should consider whether we need to. Whether, that is, we can construe science in such a way that it has no need of induction. This is the approach famously developed by Sir Karl Popper. Popper agreed with the sceptic that induction cannot play a part in rational thought. But he denied that this means that we must be irrationalists about science.

Since he was a rationalist about science, Popper was obliged to supply an alternative account of scientific inference to replace induction. This must allow scientists to decide between competing theories and guide them in their method. The alternative must be purely deductive, as only deduction is necessarily truth preserving and does not "go beyond the evidence" as induction does. The alternative that Popper proposed to the inductive verification of theories was the deductive falsification of them.

As was maintained by philosophers well before Hume, because theories usually involve generalizations over an infinite number of cases, they can never be verified directly. However many consequences have been shown to be true, there will always be consequences of the theory that have not been tested and hence may be false. On the other hand, if one consequence has been shown to be false, then it follows that the whole theory is false. Thus we cannot know theories to be true. We can know them to be false, or we can know that as tested so far a theory has not been shown to be false. The scientist will reject the falsified theories but will retain the unfalsified ones; the latter will be "tentatively accepted".

This is the basis of Popper's account of the epistemology of science. Our knowledge is either negative (we know that a theory is false because we have falsified it) or positive, in which case it is merely conjectural. That is, an unfalsified hypothesis which we are testing is currently conjectured to be true or near to the truth. We subject the hypothesis to testing not in order to gather evidence in favour of its truth, because the whole point of anti-inductivism is that there is no such thing as evidence for the truth of a generalization. Rather, the point of testing is to try and falsify a hypothesis. So the whole process can be summed up thus: we make conjectures and then try to refute them. Those which have not been refuted have the status, temporarily perhaps, of conjectural knowledge.

It is part of Popper's rationalism that he will claim that theories that have withstood stringent testing are in some sense better, or more likely to be "better" or nearer the truth, than theories that have been falsified or have not been subjected to such severe testing. We shall see that this rationalist aspiration is inconsistent with his rejection of inductive reasoning.

Although Popper is keen to save rationality for science, it is worth pausing to see how far his position diverges, even on his own admission, from what one might ordinarily expect from someone who regards science as rational. For a start, no scientist knows any theory to be true. Consequently, for all the endeavours of recent decades and centuries we have no more positive scientific knowledge than did the contemporaries of Galileo. Nor have all the experiments and observations made during this time added anything to the likelihood of any theory. Our most successful theories have had no confirmation whatsoever. This means they have been confirmed just as well as our least successful theories. Of course our least successful theory may have been falsified already. And our most favoured theory may not yet have been falsified. But for an anti-inductivist can that be a reason to think that the most favoured will remain unfalsified?

Consequently, Popper's notions of science, theory, and hypothesis are highly restrictive. For a claim to count as scientific it must be potentially falsifiable. The criterion of falsifiability is not one which scientists themselves have observed. I mentioned in the Introduction that Darwin took his views to be confirmed by fossil remains, while believing that failure to find the relevant evidence would, because of the nature of the fossil record, not count against him. Darwin's hypothesis has this relationship with the evidence because it is a complex explanatory hypothesis, not simply a generalization. Consequently, data that his theory would neatly explain confirm it, while the absence of the same data need not harm it. Popper's philosophy is thoroughly Humean, not only in its rejection of induction, but also in its conception of hypotheses as Humean generalizations.

Other kinds of hypothesis fail to register as scientific by Popper's criteria. Any statistical hypothesis fails. Consider Mendel's hypothesis that the ratio of plants showing dominant traits to those showing recessive traits should be three to one. Imagine he had found after a thousand trials that 97 per cent of plants showed the dominant trait. Is his hypothesis falsified? If you think not, then what about a million trials and a proportion of 99.9 per cent? The point is that neither of these results nor any other possible result is logically inconsistent with Mendel's hypothesis. It is always logically possible that future trials will go the other way so that the ratio ends up where it ought to be, and even if they do not, that outcome is compatible with the ratio of dominants to recessives in all the infinitely many possible trials being three to one. In Popper's sense of the word "falsify", whereby an observation statement falsifies a hypothesis only by being logically inconsistent with it, nothing can ever falsify a probabilistic or statistical hypothesis. Hence it is unscientific.

What does Popper say about this? Popper's response is that falsifiability is a criterion not of statements or theories alone, but also of those things plus the method used to evaluate them. As we shall see in Chapter 8, it is difficult to say much that is *a priori* about method. Nonetheless, Popper remarks rightly enough that "the physicist knows well enough when to regard a probability assumption as falsified". But of course the physicist knows this only because he uses induction, reasoning that the proportion displayed in a large sample is very likely to be similar to that existing in the population as a whole. Popper himself seems to suggest that we regard it as a *convention* that hypotheses like the Mendelian one would be falsified by samples like the ones mentioned. For a start it is not clear how a convention could overturn a principle of logic. If A and B are consistent, how can a convention make them inconsistent? To be charitable to Popper we may consider his proposal as an extension to the meaning of "falsifies". A theory is scientific if it is falsifiable – in the extended sense which covers both the logical cases of falsification and the cases of probabilistic divergence.

What Popper has done is to allow inductive reasoning into the story under the guise of convention. If we can do this for "falsifies", why not for "scientific", "confirms", and all the other vocabulary of science? One example flows directly from the convention already mentioned. Say a thousand trials show a proportion of 74.881 per cent with dominant traits. By Popper's convention we can say this falsifies the hypothesis that the proportion is greater than, say, 80 per cent, and also the hypothesis that the proportion is less than 70 per cent. Since these two hypotheses are falsified, we are left with the knowledge that the proportion lies between 70 per cent and 80 per cent. Thus we have confirmed a relative of Mendel's hypothesis, the hypothesis that the proportion is 75 ± 5 per cent. This confirmation is no more inductive than Popper's conventional notion of falsification. But as the confirmation clearly is inductive, it must be that the notion of falsification is inductive too. Because it is inductive we can use it to fashion, within Popper's system, all sorts of inductive notions. This is a feature of Popper's philosophy – when the going gets tough, induction is quietly called upon to help out.

Let us now move on to Popper's way of spelling out the idea that one theory can be objectively better than another. By "objectively" I mean to refer to their relation to reality, independent of any testing of them. Popper makes use of the notion of *verisimilitude*, which we have already encountered. The idea here is that one theory can be nearer to the truth even if neither is bang on correct. There is no doubt that this is an important notion, although it is difficult to make it precise. We will look at Popper's (unsuccessful) account of verisimilitude in Chapter 8. Its details need not bother us here, for the central difficulties with Popper's anti-inductivism remain whatever notion of verisimilitude we employ. These problems arise when we begin to ask: What sort of empirical results could be taken to suggest greater verisimilitude in one theory than in another?

Popper points to the *corroboration* of a theory, which is, loosely, an evaluation of a theory's past performance, specifically "with respect to the way it solves its problems; its degree of testability; the severity of tests it has undergone; and the way it has stood up to those tests". Corroboration, emphasizes Popper, is purely backward looking. It is only the history of the theory in observed circumstances that goes into its corroboration. In contrast to an inductivist's view of past performance, corroboration is not to be regarded as evidence in support of a theory. Rather, it is simply a report of our past efforts to falsify the theory. Corroboration is therefore supposed to contrast with confirmation.

Nonetheless, Popper says we should prefer a well-corroborated to a poorly corroborated theory. It sometimes seems that this amounts to a definition of preferability for Popper. But, assuming it is not, can the anti-inductivist give genuine grounds for preferring one theory to another? Perhaps one of two competing theories has been falsified and the other has not. This at least gives some reason for preferring one theory, for at least we know the falsified theory to be false. The possibility is still left open that the other is true. As I have already suggested, past immunity to falsification cannot be used to make any inference about the future success of a hypothesis against testing, unless one argues inductively. Nor, for that matter, can the falsification of a theory be a reason for an anti-inductivist to think that it can be falsified again. So the anti-inductivist is not entitled to think that the falsified theory will perform worse than the unfalsified theory in new tests or tests that we might have carried out but did not.

This point generalizes to the cases where both theories may have been falsified or neither of them has. What we want is for a method of comparison to leave us with theories of higher verisimilitude than the theories we have discarded. This is what grounds for preference should amount to. Does corroboration provide such grounds for preference?

Popper says of the practical person – one who has to use a theory as a basis for action – that they should not rely on any theory, because we have no reason to think any theory is true. But it is rational to choose to operate with the better corroborated theory. Popper thinks this is just obvious. Either this is because it is part of the definition of "rational" to prefer a well-tested theory over a poorly tested theory, or it is because we regard better tested theories as being more likely to be true or to have greater verisimilitude.

The first of these alternatives bears a resemblance to Strawson's views, which we will come to shortly. It will not do. Although it is pleasing to be called rational, this is not what the practical person is primarily interested in. This person is about to act and has at least two courses of action open, each of which might yield the desired result. Which course of action will actually yield the desired result will depend on the truth of the various theories under consideration as they apply to the (future) occasion of this person's acting. What the agent wants is to have good reasons for believing that one course of action is more likely to be successful than the alternatives. The rationality of the choice is really very little to the point – unless present rationality is connected to future success. To the practical person, success tomorrow is of more value than being rational today. And with regard to the past performance of various theories, their degrees of corroboration will only be a guide to their future success if some form of inductive reasoning is acceptable.

So let us return to our question: Does corroboration provide such grounds for preference, where "preference" means preference with regard to truth or verisimilitude? The answer is that it cannot do so if one rejects all forms of inductive argument. Or, to put it another way, only if some form of inductive argument is valid can corroboration be any indicator of verisimilitude. For corroboration concerns only past observations, while verisimilitude concerns not only past observations of a theory's performance but also its past unobserved performance and future performance, both observed and unobserved. Corroboration will tell us directly only about a small, indeed infinitesimally small, part of a theory's total performance. So for it to tell us something about the majority of a theory's performance, i.e. the bulk of its verisimilitude, then we will have to make inferences from the observed to unobserved and from past to future. But no such inference can be acceptable to the anti-inductivist.

Our conclusion must be then, that whatever the difficulties posed by Hume's problem, they must be overcome if we are to have a rationalist account of science. We have considered the thought that we might be able to give an account of science that does not employ induction. We have found that, after all, this is a dead end. So either science is irrational, or it must be that the use of induction, despite the claims of the sceptic, is unobjectionable. (I will say more about falsificationism and Popper's philosophy of science in Chapter 8.)

Dissolving the problem of induction

We saw that the attempts to justify induction lead to circularity or regress of ever more general propositions and inferences about the unobserved. One escape route might be to suggest that there may be some form of reasoning in this chain which does not need any independent justification. Perhaps we can start from a point at which the request for further justification is somehow illegitimate or redundant.

This view is taken by Sir Peter Strawson. He says that it is a confusion to ask what reason we have to place reliance on inductive procedures. His argument is analogous to an argument that purports to show that it is confused to ask what it is that makes deductive reasoning valid. He first points out that it is only individual instances of deductive reasoning or particular forms of deductive reasoning that can rightly be called valid. To ask what makes deductive arguments in general valid is to ask something senseless. For deductive arguments are not valid in general. Deductive arguments are often invalid. Someone may make a mistake or commit a logical fallacy. So the appropriate question is: What makes a particular deductive argument valid? Strawson's answer is that this is a matter of applying deductive standards. And whether something, e.g. *modus ponens*, counts as a deductive standard is analytic, that is to say, is simply a matter of the meanings of the words concerned.

Similarly, it is only individual inductive arguments that can be assessed, not inductive reasoning in general. In the case of inductive arguments it is not validity for which they are assessed, but reasonableness. On the basis of certain evidence it may be reasonable to make a certain claim or it may not. It will be reasonable insofar as the evidence strongly supports the conclusion, and unreasonable insofar as the evidence is weak. Such claims are clearly analytic. Furthermore, says Strawson, it is analytic that the strength of evidence is related to the number of favourable instances and the variety of the circumstances in which they have been found. So he says, "to ask whether it is reasonable to place reliance on inductive procedures is like asking whether it is reasonable to proportion the degree of one's convictions to the strength of the evidence. Doing this is what 'being reasonable' means in such a context."[47]

Strawson's point can be summarized as follows. Asking for the justification of induction in general is misplaced. What we can ask about is the justification of individual inductive arguments. And here justification is a matter of showing the argument to be reasonable. That, in turn, is a matter of applying the standards appropriate to inductive arguments. Those standards do not stand in need of justification, since they are what gives meaning to such phrases as "reasonable inductive argument". Just as the metre rule in Paris is a standard for something being a metre long. We do not need to justify this, for "one metre long" means the length of just that rod.

An initial problem for Strawson's view is its lack of specificity. Indeed it is *a priori* that it is reasonable to proportion one's degree of belief to the strength of evidence. But the interesting question is: When is evidence strong? Strawson mentions features that make for strong evidence, these being primarily the quantity of evidence and the variety of instances. Is this sufficient to make for strong evidence? We saw that the uniformity premise was too broad to be both true and useful, and Strawson's characterization suffers for the same reasons. We can find clear counterinstances. If I take balls out of a bag about which I know nothing, and all the ones I see are green, then I might think that each green ball I take out should strengthen the belief that the next one will be green. But, if I have additional information that there is or may be a red ball in the bag, my belief that the next ball will be green should weaken with each additional green ball removed. Is it reasonable to conclude after observing the colour of the leaves on New England's maple trees for a summer that they are always green? It might be reasonable for a visitor from Kenya, but not for one from Ireland. Does the repetitive strategy of a poker player show that he will continue to repeat it? If you think he is a novice, then it might; if you think he is experienced, you might instead expect his strategy to change at any moment.

The point of these examples is that the reasonableness of an inference cannot be determined in isolation from background knowledge. I urged the same point in connection with the hypothetico-deductive model of confirmation and with Goodman's problem. All three focus on the same question: when is evidence good evidence for a hypothesis? My response was, and is, that there is no model of confirmation or of the evidential relation.

Nonetheless, Strawson can be understood to be arguing for something which is true: that it can be reasonable to make an inductive inference, and correspondingly unreasonable to reject it, and this is dependent on one's total evidence (everything one knows). In Chapters 1 and 2 we saw that it is part of the concept of law and explanation that certain facts count as good evidence for the existence of the law or for the truth of the explanation. The corresponding inferences therefore count as rational. My view is that the connection between evidence and law or explanation being *criterial* explains the respect in which Strawson is right. If we have undefeated criteria for P then it is rational to infer P. It is because this criterial relationship is holistic that there can be no model or template for the application of any of these concepts.

Yet I also denied that this is a solution to the problem of induction. An inference may be very reasonable and yet fail. The criteria may all be fulfilled and undefeated and there be no law. The criterial relation is not deductive. Take Ohm's law as applied to copper. It states that for low currents in a wire and potential differences across the wire the ratio between them is constant. The relevant criteria are well satisfied. There are countless comfirming instances of the law under direct testing. It is simple. It integrates well with other laws to explain a variety of phenomena concerned with electrical circuits. It can itself be derived from more fundamental laws of electrodynamics for which there is even broader evidence. Given this evidence it is reasonable to believe that Ohm's law is indeed a law. This fact is part of the concept of law; it is *a priori*. Nonetheless, it is still possible that there is no law here. The question is then, even if it is rational to believe that there is a law, does that fact make someone who has that belief more likely to be right than someone who believes there is no law? As we saw with Popper, earning the right to be called rational is of little interest unless some connection is established between rationality and success.

Part of what is going on is an ambiguity over what counts as the assessment of an inductive argument. Strawson says that we assess arguments for their reasonableness. But we can also assess arguments for their propensity to lead to the truth. Hume's argument focuses on the latter, and for good reason, because the point of inference is to acquire true or nearly true beliefs. What we need to know from Strawson is whether an assessment in terms of reasonableness has any bearing on its assessment for propensity to lead to the truth. But Strawson says nothing that might establish such a connection.

Another way to see this point is to imagine that there were other standards to apply to inductive arguments, standards which say whether the argument is more or less dotty. For instance, an argument may be dotty where the evidence seems completely unrelated to the hypothesis. So we can assess arguments for reasonableness and dottiness. Bearing in mind the aim of truth, the question is why we should adopt reasonable arguments and not dotty ones. Does someone who argues reasonably stand a better chance of getting things right than someone who argues dottily? Of course the answer is yes, they do; but as soon as we try to explain why, we are back with Hume's problem. The core of Hume's problem is not so much about whether it is intellectually respectable to argue inductively, but whether one can show that one is more likely to be successful (in getting to the truth) in so doing.

Further reading

W.V. Quine introduces the Duhem–Quine thesis in "Two dogmas of empiricism". N.R. Hanson's *Patterns of discovery* is a classic on theory and observation. Sir Karl Popper's *The logic of scientific discovery*, among other of his writings, presents the falsificationist approach to science. Sir Peter Strawson's dissolution of the problem of induction is to be found in his *Introduction to logical theory*.

Chapter 6

Probability and scientific inference

A probabilistic justification of induction?

Statistical and probabilistic reasoning plays an important role in scientific inference. It is sometimes thought that such reasoning provides an answer to the problem of induction. While induction cannot give us logical certainty, because we cannot logically rule out alternatives, it might be that it can lend a high probability to its conclusions. In this chapter, as well as describing some of the probabilistic–statistical techniques used in science, I shall examine in some detail the hope for an *a priori* probabilistic inductive method. Before starting, it should be clear that the prospects are not very good as regards a thoroughgoing response to Hume's problem. In general, Hume's problem asks: How can observations limited to a small region of space–time tell us what is happening throughout space–time? The conclusion of an inductive argument, whether it be an argument by naive induction or sophisticated statistical reasoning, will always reach out beyond the evidence. There will always be room for the Humean sceptic to respond that any method we employ rests upon some assumption to the effect that the observable portion of space–time is some sort of reflection of the universe as a whole. As the rest of reality might be utterly unlike the observed part, this principle is clearly empirical and so cannot be justified by any *a priori* (e.g. mathematical) reasoning.

It is nonetheless instructive to see precisely what sort of difficulties such forms of inductive reasoning run into. In particular, the difficulties illustrate that these methods make assumptions rather stronger and more specific than a general assumption such as "nature is uniform". For instance, if I want to make an inference about the whole of a certain population on the basis of a sample, I need to assume not only that there are laws governing the phenomena I am interested in, but also that my sample is a fair representation of the whole population. The latter is a stronger assumption, as there could be laws governing this population (e.g. laws governing the mortality of humans), but my sample, by sheer bad luck, might be highly unrepresentative. So the assumption that my sample is representative is a claim additional to the assumption of the existence of laws.

Kinds of probability

Subjective probability

Before we get into the details of probabilistic reasoning, we need to ask what is meant by "probability". There are two sorts of answer to this question, which go under the descriptions "objective" and "subjective". Subjective probability measures a person's strength of belief in the truth of a proposition. Objective probability concerns the chance a certain sort of event has of happening, independently of whether anyone thinks it is likely to occur or not.

Probabilities are measured on a scale of 0 to 1, where 1 represents certainty that the proposition is true and 0 certainty that it is false. Consider a coin that has been tossed. You are asked what probability you attach to the proposition that heads is uppermost. If you know nothing about the situation, you would, I expect, respond with 0.5. This is because there are two possibilities and you have no more reason to think that one is actually the case than the other. One way to think of the matter is in terms of the odds you would think fair if you were asked to bet for or against the proposition. In bookie's terms, odds are expressed as a ratio, for instance seven-to-one. The ratio is between the sum the bookie will pay you if you win and the sum you pay the bookie if you lose (i.e. if you bet on a horse at odds of seven-to-one, 7:1, you win £7 if the horse wins and lose £1 if it loses). In the case of the tossed coin the odds you should think fair are 1:1 (known as "evens"); you win £1 if heads are shown and lose £1 if tails are revealed. Given the probability you attach to heads (0.5) the amount you expect to win is zero, since there is 0.5 probability of winning £1 and a 0.5 chance of losing £1. (The amount you stand to lose is your *stake*.) The amount you expect to win, taking account of your subjective probabilities in this way, is the *expected value* of the bet. A bet at *fair odds* is one where the expected value is zero. If the odds were higher than 1:1 then the expected value would be positive and the bet would be favourable – you expect to gain from it. If the odds were lower than even then the expected value would be negative – you expect to lose money. Bookies make bets at odds that are favourable to themselves.

Now imagine that you learn some further facts, for instance that the coin had been tossed by a cheat who stands to win if tails are revealed, and that of the previous ten tosses eight have yielded tails. You might now be inclined to change your willingness to bet equally on heads as on tails; instead you are more strongly inclined to bet on tails. Alternatively, the odds would have to be raised so that you would win a lot for a small stake before you would be willing to bet on heads. Reflecting on the previous ten tosses, you might be four times more confident that tails will show than heads, i.e. the probability of heads is 0.2 and the probability of tails is 0.8. To be induced to bet on tails you would have to be offered odds of at least 4:1. At 4:1, the expected value of the bet is $(£4 \times 0.2) - £1 \times 0.8) = £0$. Any poorer odds and you would expect to lose money.

One way of eliciting subjective probabilities – or, what amounts to the same thing, your estimation of fair odds – is as follows. You are forced to take part in the bet. You do not know which side of the bet you will be asked to take. But you are allowed to choose the odds. As you do not know whether you will be asked to bet on or against the proposition, you need to choose odds so that the expected value of the bet is zero for both sides. That way you will be equally happy whichever side you are forced to bet on. In the example of the biased coin, your fair odds are 4 : 1 – at lower odds the bookie would force you to bet on heads and you would expect to lose, and at higher odds he would force you to bet on tails, and again you would expect to lose.

Objective probability and chance

Contrasting with this subjective probability is the notion of objective probability. Here the idea is that the world itself contains objective chances or probabilities, to be understood independently of notions such as evidence, belief, confidence, etc. Let us return to our coin-tossing case. We like to think that a normal coin is as likely to land heads up as tails up. This fact tends to display itself in a long run of tosses, which will contain roughly equal numbers of both outcomes. But reflecting on the history of four times as many tails as heads in our case, as well as on the dubious character of the coin tosser, we might well surmise that the coin is not an ordinary one. Perhaps it has been fixed so that it is more likely to fall tails up – for example by making one face of the coin out of a denser metal than the other. Such a coin is said to be *biased*. The objective probability of tails is a measure of the bias of the coin. The bias and the probability are objective features of the coin, like its mass and shape. These properties have nothing to do with our beliefs about the coin.

Objective probabilistic properties also go by the name of *chance*. The fair coin has a chance of landing on heads of 0.5, while the biased coin has a chance of landing on heads that is less or more than this. While this is intuitively easily understood, defining the concept of chance is more difficult. We have already encountered the problem in connection with probabilistic laws. What is it for a coin to have a chance of 0.2 of landing on heads or for a nucleus to have a chance of 0.3 of decaying within the next minute? One proposal, the minimalist's, is that the chance of heads is equal to the proportion of tosses that land on heads in a long run of tosses; similarly, the chance that a certain kind of nucleus decays within a given period is equal to the proportion of nuclei that actually do decay within that period.

The objection to this is our intuition that it is possible for the chance to differ from the long-run proportion. A coin may be a fair, unbiased one. But it is still possible for a long sequence of tosses of the coin to show more heads than tails. It is unlikely that the proportion of heads will in the long run be very different from 0.5, but it is nonetheless possible. According to this objection, the objective chance is not *identical* to the long-run frequency. Rather the chance is that property which *explains* why a long sequence of tosses tends to a proportion of heads near to 0.5. A better account of the relation is provided by the criterial view we saw in Chapter 1. The observed long-run frequency will be criterial for the law. And so, other things being equal, our best estimate of the probability of the event will be its frequency. But it is not necessary that the long-run frequency and the estimate are the same. So the intuition is satisfied, while at the same time giving an explanation of what the objective, law-given probability of an event is.

The notion of objective chance is not accepted by all. In the tossed-coin example, when asked to assess the probability of heads, we attached the probability 0.5 out of ignorance – we have no more, or less, reason to believe that the coin will land on heads than it will tails. But it is in fact already either heads or tails, and there is an objective sense in which if it is heads it stands no chance of becoming tails. One might think that something similar is true even before the coin has landed. As the coin is flying through the air one might attach a subjective probability of 0.5 to its landing on heads; but one might also think that the various laws governing the coin's motion are deterministic, that given the weight, shape, and velocity of the coin, as well as relevant facts about the air through which it passes and the table on which it lands, it will inevitably land on heads (or tails as the case may be) with the same inevitability that a goose's egg falling 15 floors onto York stone paving will smash. Someone who knew all the relevant facts and could calculate with amazing rapidity might be able to figure out which side up the coin will land. This reflection inclines some people to think that in the deterministic case there is no real objective chance – the chance is only subjective, because it is a reflection of our ignorance.

However, contrasting with this is the indeterminacy of certain *irreducibly* probabilistic laws, such as the laws governing the decay of atomic nuclei. These are irreducible in the sense that there are no further facts or laws that altogether fix deterministically the decay of a particular nucleus (which would make nuclear decay like coin tossing). (Some physicists have believed that nuclear decay is like coin tossing. Such a belief prompted Einstein's famous remark that "God does not play dice with the universe". According to Einstein's view there are hidden variables that do determine decay. It is our ignorance of these that makes decay look probabilistic. But most physicists accept that Einstein was wrong about this.) If there are irreducibly probabilistic laws then the chance that a nucleus will decay in a certain time will be an objective property of that nucleus. Irreducibly probabilistic laws put the case for an objective notion of probability in its strongest form. But many proponents of such a notion want to attach it to deterministic cases too. The construction of a coin makes a difference to the frequency of heads and tails – we can make biased coins. As the construction of the coin is an objective feature of the coin, and as it affects the long-run frequency, it follows that there is an objective feature that explains the frequency of heads in a long run of tosses. If we interpret the notion of objective chance as the possession of an objective property which explains long-run frequencies, then we are entitled to say that the coin possesses an objective chance of landing heads, even if every toss is a deterministic event.

There are debates about which of the two notions of probability, subjective and objective, is preferable. It seems to me that both notions have their place. On the one hand it seems that we do employ a notion of subjective probability, because we attach probabilities to events to reflect our state of evidence or ignorance, for instance in cases where the event in question, like the tossed coin, has already occurred. On the other hand, the scientific evidence is that Einstein was wrong about indeterminism, and that there are irreducibly probabilistic events. Furthermore, there seems to be good reason to attribute deterministic tendencies to things such as coins, dice, buttered toast, catching a cold, and dying young. Here the probabilities are objective – they appear to be properties that can be discovered and used for predictions. One approach which seeks to combine these two notions is the *personalist* conception of probability. The personalist conception agrees with the subjective view that individuals have degrees of belief and also with the objective view that there are objective probabilities. According to the personalist a rational individual will seek to match his or her personal degrees of belief to the relevant objective probabilities.

Classical statistical reasoning

Some populations are what is called *normally distributed*. The normal distribution is characterized by the familiar bell-shaped curve shown in Figure 6.1. Many properties of living creatures are approximately normally distributed, e.g. the height of trees in a wood, the weight of honey carried by bees in a hive, and the number of eggs laid by chickens in a year. It is because the bell curve continues infinitely in both directions that the distributions of these properties are only approximately normal – one cannot have trees with negative height, or a chicken that lays a million eggs in a year. Nonetheless, these approximations are found useful by statisticians, and we shall now look at the sorts of reasoning they employ.

There are two parameters needed to describe a normal distribution, the *mean* and the *variance*. The mean is just the average value of the variable in question. The mean is found at the centre of the curve, where it is also at its highest (μ in Fig. 6.1). The variance of such population is a measure of the flatness of the bell curve. A population with a small variance is concentrated around its mean, while a population with a large variance is more spread out. Instead of talking about the variance, it is often more convenient to refer to the *standard deviation* of a population (σ in Fig. 6.1), which is the square-root of the variance. In a normal population 95 per cent of the population falls within 1.96 standard deviations either side of the mean.

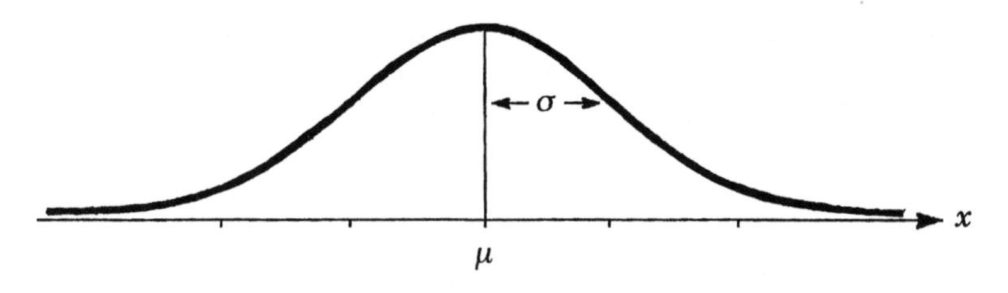

Figure 6.1

I will consider one probabilistic–statistical method that, mathematically, seems to hold out some promise of inductive success. The inductive problem to which this method aims to provide a solution is this. We have a sample taken from a population which we know to be normally distributed and the variance of which is known. What is the population mean? For instance, we want to know the average height of the spruce trees in the Tay Forest. But we cannot measure all the trees, so we take a sample of spruce trees and find the average for this sample. What can we infer about the average height of all the spruce trees in the forest?

In stating this problem we have already assumed a certain amount of knowledge, which typically could only be gained inductively. To use mathematical reasoning based on the normal distribution we have to assume that the heights of these trees are approximately normally distributed. (At best this can only be an approximation, for the reasons mentioned above, and because the total population is finite.) Such an assumption could only be justified by inductive means, whether generalizing from a sample of spruce trees or by inference from facts about trees, which could only be known inductively.

Estimating the mean height of all the spruce trees in the forest on the basis of a sample is a further inductive leap. It is this induction which we will examine. If the population is much larger than the sample, then the two could differ to a very great degree. If the average height of all the trees is 10 m, it is not logically impossible that the mean height of a sample of 20 spruce trees is 14 m. Nonetheless, so the statistical argument goes, it is extremely unlikely.

Figure 6.1 represents the population in which we are interested. The variable quantity x (the height of a tree) has a mean μ and a variance σ^2 (and a standard deviation σ). Consider all the possible samples from our population, each sample being a group of n spruce trees from the Tay Forest. Each sample has its own mean. We can regard these sample means as themselves forming a population. Let us call the new variable, the mean of a sample, y (Fig. 6.2). This population of sample means is related to the original population of tree heights. The means of the two populations will be the same – the average tree height is equal to the average of all the average heights of all the groups of n trees. The variance of the new population is smaller than the variance of the original population. The larger the size of the sample n, the smaller the variance of the sample means. One way to see this is to think that the larger the sample the nearer it is to being the whole population, and so the nearer the sample mean is likely to be to the population mean. The new variance σ^{*2} and the original variance σ^2 are related thus: $\sigma^{*2} = \sigma^2/n$ (and so the new and old standard deviations are related thus: $\sigma^* = \sigma/\sqrt{n}$). (Note that if $n = 1$ the samples are single trees, and the population of the sample means is just the population of tree heights, so $\sigma^{*2} = \sigma^2$. If $n = \infty$ then there is only one sample, the whole population. In which case the sample mean is identical to the population mean, μ. So $\sigma^{*2} = 0$. Remember, we are making an idealized assumption that the population is infinite.)

Let Figure 6.2 represent the distribution of the means y of the possible samples. We can then say that the probability p that a random sample will have a mean falling within an interval (a, b) will be proportional to the area under the curve bounded by a and b, as shown in Figure 6.3. We can express this as:

$$p(a < y < b) = \text{Shaded area}/\text{Total Area}$$

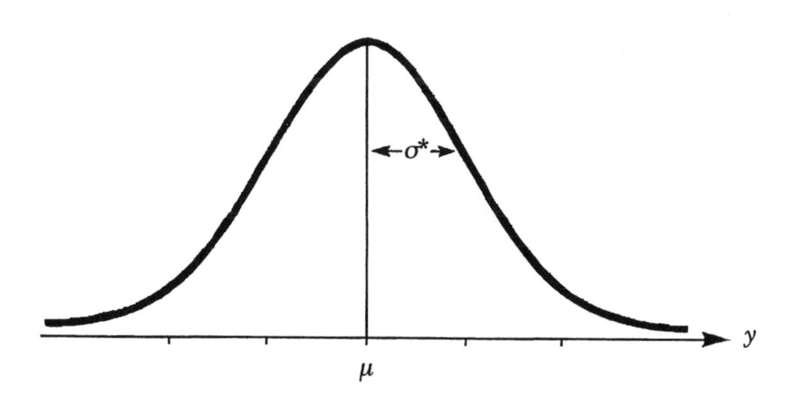

Figure 6.2

For given values of μ, σ^*, a, and b, the shaded area can be calculated. If a and b are 1.96 σ^* either side of μ, then the area is 95 per cent of the total. Thus a random sample has a 95 per cent chance of falling within these limits. Such limits are called *confidence limits*; in this case they are 95 per cent confidence limits.

This much is the theory based on a given mean μ. But μ is what we want to discover. So we take a sample of n trees and find that its mean is c. From the theory just discussed we know that $p(c$ lies within the limits $\mu \pm 1.96\ \sigma^*) = 0.95$. The value of c is the sample mean, which we know, and μ is the population mean, which we do not know. We infer from this that $p(\mu$ lies within the limits $c \pm 1.96\ \sigma^*)$ is also equal to 0.95. This is illustrated in Figure 6.4.

It can be seen from the figure that 95 per cent of the sample means fall within the shaded area. For each such sample mean, μ falls within 1.96 σ^* of it – in Figure 6.4, c falls within 1.96 σ^* of μ, and therefore μ is within 1.96 σ^* of c. This suggests that the probability that μ falls within 1.96 σ^* of a randomly chosen sample mean is 0.95. Hence the inference mentioned above. It looks as if we have a probabilistic inductive result based on purely mathematical reasoning. From the sample we have inferred an interval such that the chance that the population means falls within it is 0.95, i.e. 0.95 is the *confidence level*. We could have found intervals for which this chance, the confidence level, is

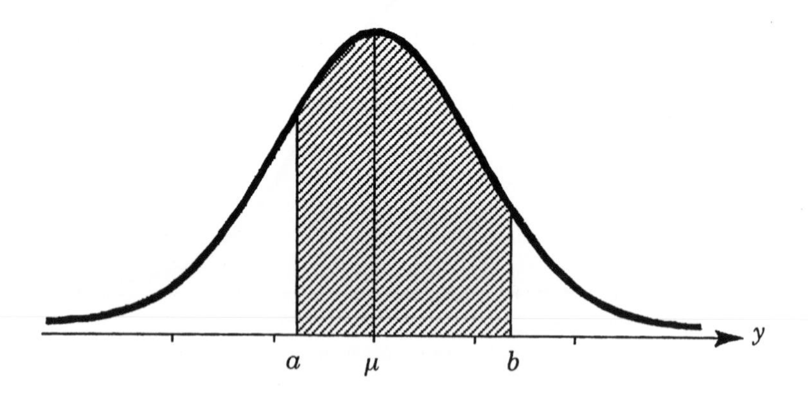

Figure 6.3

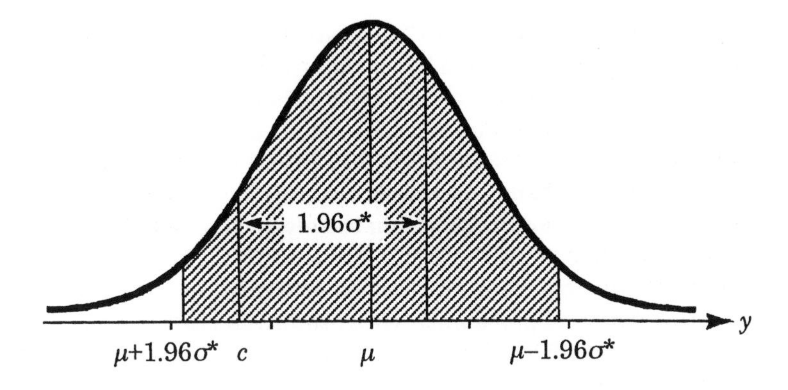

Figure 6.4

greater or smaller, but note that we could not specify a finite interval within which μ falls with certainty; the only such interval is $c \pm \infty$. Nor can we specify a precise value for μ with non-zero probability. At a given confidence level, the larger the sample, the narrower the interval, since σ^* gets smaller as n gets bigger (remember that $\sigma^* = \sigma/\sqrt{n}$). So with more evidence we can be more specific about the value of μ but never exactly precise, except in the trivial case when the sample is identical with the population ($n = \infty$).

We have a mathematical method of induction which accords with what we have said about induction, i.e. it cannot give certainty, but nonetheless gives useful inductive results, allowing us to say with a high degree of probability that the value of a certain variable falls within a certain interval.

Is this a solution to the problem of induction? You will recall that we have already admitted that this all rests upon facts that could only be known inductively – that the population is normally distributed with variance σ^2. But there are problems beyond this particular prior use of induction. We inferred from the fact that 95 per cent of n-membered samples fell within a given area that the chance of a particular n-membered sample falling within that area is 0.95. This inference is legitimate only if each possible sample has an equal chance of being chosen.

What does it mean to say "each possible sample has an equal chance of being chosen" and hence that "the probability that μ falls in the interval $c \pm 1.96\,\sigma^*$ is 0.95"? The theory of confidence intervals is most closely associated with an objective notion of probability. On the face of it we may prefer the objective notion, because we want our conclusion to provide an objectively high chance that the mean falls within the given confidence limits. We want our inductive methods to be reliable, but if they cannot be absolutely reliable, then let them be nearly so. In the light of an objective notion of probability, let us return to the key premise of the theory of confidence intervals, that each possible n-membered sample has an equal chance of being chosen. In the objective sense of "chance" this premise may be true, but it may not be. That is a question which may need to be answered on the basis of independent inductive reasoning; in particular, this is the case where there are constraints on the choice of sample. For example, I may wish to estimate the average weight of the human brain. Let us suppose that brain scans and other indirect ways of measuring brain weight are unsatisfactory for my purposes. So the only way of getting my data is by weighing individual human brains. I cannot legally, or even ethically, do this by selecting at random victims to have their brains removed and weighed. But I can measure the brains of the recently dead. This means that the chances of my sample being chosen are manifestly higher than any possible sample, since my sample is restricted to the recently dead while the population includes living people too. This does not invalidate entirely the use of statistical reasoning. But it does mean that we have to think carefully about how representative the sample is of the overall population. Of course, the sample is completely unrepresentative in an important respect. What we must do is consider the respects in which this unrepresentativeness may be relevant to the variable being estimated. So in this case we would need to take into consideration the fact that the recently dead tend to be overrepresentative of the older portions of society and that they are more likely to have been diseased or unhealthy at the time of death. These factors may well influence brain size, and so the data need to be adjusted accordingly.

This means that the application of the theory of confidence intervals can only be made with the assistance of further knowledge, which itself will be gained inductively, and so the use of the theory will not be *a priori*. Furthermore, the sort of reasoning referred to above, along with the adjustment of the data, do not serve to show that the chance of choosing the sample was the same as that of choosing any other sample. What it does is to remove any known unrepresentativeness, which is a different matter altogether. In which case it is not the classical theory of confidence intervals that we are using, but something different. Note that this additional reasoning is providing us with premises that have a role similar to the uniformity premise (or its relatives) in standard Humean inductions – premises which tell us that the whole population is relevantly similar to our sample.

Furthermore, it should have seemed odd all along that the chance of choosing that particular sample should come into the picture in the way the classical theory of confidence intervals suggests it does. Much seems to hang on the method by which the sample was chosen, while what really matters is the relationship between the sample and the various possible values of the population mean. Of course, biases in choosing the sample are important in assessing that relationship, but not every difference in the chance of choosing a sample is the result of a relevant bias. (For instance, the unavailability of scales for weighing brains over 10 lb would mean that samples including vastly overweight brains stood no chance of being chosen. But that would not affect the inferences I should draw from my sample if in fact all the brains in my sample weighed less than this.)

In addition, in the classical view the way in which we *describe* the method by which the sample is chosen has a significant bearing upon the conclusions we may draw from the sample. This suggests that a strongly non-objective element is present in the use of the theory. Let us say that in estimating the mean of the population I decided not to choose a sample of some predetermined size, but to let the size depend upon the role of a die (e.g. Sample size $= n \times$ number on the die, for a fixed value of n). We could calculate a distribution curve for the means of samples thus obtained, but it would not be the same as in the curve in Figure 6.2, which shows the distribution for all n-membered samples. As the new rule for choosing the sample size allows n as the size of the smallest sample but also allows larger samples, then the new distribution will be a rather sharper, narrower curve (hereafter called the "new distribution"), as shown in Figure 6.5.

As the new distribution is narrower, a larger proportion of it will fall within a given interval, and so the confidence level will correspondingly be greater. Consider two statisticians, Bill and Ted. Ted has decided that his sample size will be n and so operates with the old distribution in Figure 6.2. Bill will choose his sample size by throwing the die, and so operates with the new distribution. Imagine that the roll of the die yields 1 and so determines that Bill chooses a sample of size n. So Bill and Ted choose samples of precisely the same size – perhaps the very same sample. As his curve is narrower, Bill will attach a greater level of confidence to a certain conclusion than will Ted. Thus Bill and Ted may draw quite different conclusions from precisely the same evidence. What makes the difference is merely how they decided to choose their sample size. This is surely an absurd conclusion. While I do not suppose that many scientists use a die to determine the sample size, I do expect that in many cases they do not fix their sample size in advance of collecting it. For obvious reasons we want as large a sample as possible. But this will be constrained by factors such as time, cost, and ease of obtaining data, so we may say something like

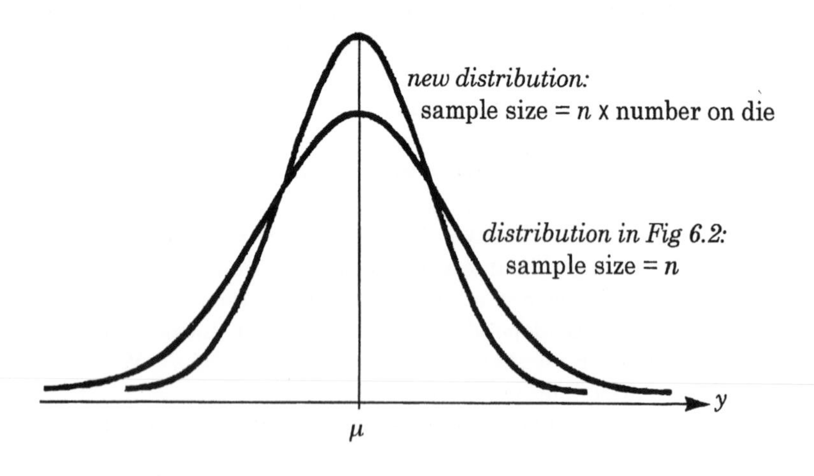

Figure 6.5

"We will gather as much data as we can in three days". If we do, we should calculate the distribution of sample means for all possible samples that might have been gathered in a three-day experiment. This in turn will require estimating, for instance, the chance of the weather turning bad and terminating our field trip early. Even if we could draw the appropriate distribution, it is strange that the conclusions we draw from it will be different from the conclusions we would have drawn had we fixed the sample size at 300 in advance, even if 300 is the sample size we in fact collected in the three days. One could imagine two researchers on the same team, one determined to finish after three days the other determined to collect 300 samples, and it turning out that the three hundredth sample is collected just at the end of the third day. One would have thought that this coincidence would have spared the two researchers any disagreement; but no, according to the theory of confidence intervals their different intentions will lead them to fall out over the breadth of the 95 per cent confidence interval they may infer from the data.

An application of similar statistical reasoning is to be found in the practice of clinical trials. Clinical trials are studies carried out to test the efficacy of new drugs, procedures, and other treatments. Typically they involve two groups, a treatment group (which receives the drug) and a control group (which does not). At the end of the trial the two groups are compared with regard to effect. Did a greater proportion die or recover in one group than in the other? One would expect there to be some differences just as a matter of chance, people not being identical. The statistical question is whether the difference is large enough to be considered significant. The question is typically framed this way: Are the results consistent with the hypothesis that the two groups are drawn from the same population? This hypothesis is called the *null hypothesis*. It effectively states that the treatment has no effect – there is at most an irrelevant difference between the two groups. The commonest statistical test of this is the χ^2 (chi-squared) test. The χ^2 test asks what the chance is of finding the measured difference between the two groups, assuming the null hypothesis to be true. If the chances of getting that difference are, say, 0.05), then the researchers can say that the difference is significant at the 5 per cent level. What this is taken to mean is: if the null hypothesis was true and this sort of trial was repeated many times, only 5 per cent of those trials would yield a treatment difference as large as the one observed.

For this reason the χ^2 test and others like it are taken by some to be measures of inductive support. There are reasons for doubting this which are similar to those mentioned in the case of estimating means. I will mention one problem of practical importance. One might think that the more often different researchers come to the same result then the better that conclusion is confirmed. But that is not the case with χ^2 tests. One instance of this is to be found in the use of clinical trials to test whether aspirin has any effect on the mortality of patients suffering from myocardial infarction. Six trials were carried out covering a total of 10,000 myocardial infarction patients, of whom 1,000 died. All the six trials found that there was no statistically significant difference between the control and treatment groups. One might have thought this to be strong evidence that aspirin has no beneficial effect, and certainly better evidence for that conclusion than the result of just one trial. But in fact the reverse is true. If we lump the data from all the trials together, regarding them as one big trial, then the difference between those who received the treatment and those who did not *is* significant.[48] How can this be? The answer is that each trial showed a small difference between the groups, but in each case the difference was in favour of the treatment group. While the difference was insignificant in each trial singly, the total difference is significant. This is because a 15 per cent difference is insignificant in small samples but can be significant for larger ones. Returning to Figure 6.5, the chances of a given deviation from the mean are smaller for large samples than for small samples.

Some statisticians are reluctant to regard such tests as measuring inductive support. At the very least it is clear that, if they do provide any such measure, they do not do so in a straightforward fashion. The theory of confidence intervals also looked as if it might provide a partial solution to the problem of induction, as it seemed to give mathematical reasons for attaching a high probability to the solution to an inductive problem (that of inferring the mean of a population from a sample). This theory would in any case have provided only a partial solution, because using the theory would depend on certain assumptions about the population which would themselves have been gained inductively. This in turn puts a constraint on the sampling procedure, that each sample should have an equal chance of being chosen. Apart from being unrealistic, it will in many cases be an empirical question whether or not this is the case, and one that needs answering inductively. It is of course possible to re-draw the distribution curve for sample means to take such features of the sampling method into account; according to the classical theory this is precisely what we should do. This means that precisely the same sample might be collected by different methods and so different conclusions should be drawn from it. This consequence of the theory of confidence intervals suggests that the fact that it has useful applications (which it does) is not due to its providing a mathematical theory of induction.

Bayesianism

These and other criticisms of classical statistical reasoning are made by those who advocate *Bayesianism*. Bayesianism is a doctrine about how a person's degree of belief in a hypothesis should change in the light of evidence. It therefore reflects a subjectivist approach to probability. Bayesianism is said, by its keenest supporters, to show:

(a) how inductive practices can be rational;
(b) how inductive practices can lead to the truth;
(c) how various problems in the philosophy of science can be solved; and
(d) what the underlying structure of actual scientific reasoning is.

If Bayesianism could achieve all this it would be a major triumph. So let us see what Bayesianism is. Bayesianism's tool in this daunting task is surprisingly modest. In its simplest form Bayes' theorem is

$$p(h/e) = \frac{p(h).p(e/h)}{p(e)}$$

where $p(h)$ is the probability of the hypothesis h, $p(e)$ is the probability of the evidence e, $p(e/h)$ is the probability of the evidence, when the hypothesis is taken as given, and $p(h/e)$ is the probability of the hypothesis, when the evidence is taken as given. These are all subjective probabilities – that is, the degrees of belief we attach to the relevant propositions. The idea is that if we put values on $p(h)$, $p(e)$, and $p(e/h)$ then there is only one value we can rationally put on $p(h/e)$, i.e. the value computed by the theorem. Say we do this, and then we carry out an experiment to test h. The experiment yields e as the result. According to Bayesians we should now change our degree of belief in h from $p(h)$ to the value we computed for $p(h/e)$. The idea is that if we were rationally committed to putting a value k on the probability of the hypothesis, when the evidence e is considered as given, now the evidence e actually is given (by the experiment) we ought now to change the probability we attach to h to k, i.e. $p_{\text{after}}(h) = p_{\text{before}}(h/e)$. This is known as *conditionalization*.

We can see that Bayes' theorem has intuitive plausibility. It says that the probability of a hypothesis, given some evidence is:

(a) proportional to the probability of the hypothesis prior to obtaining the evidence;
(b) proportional to the probability of the evidence, taking the hypothesis as given; and
(c) inversely proportional to the probability of the evidence itself prior to the test.

These relationships in turn explain the following facts about inference:

(a) If we think a hypothesis is in itself likely in the light of background knowledge, then this will be taken in its favour when assessing its probability after a test.
(b) The degree to which the evidence supports a hypothesis depends on how closely linked they are. If on the assumption that the hypothesis is true it would be very likely that we get the evidence, then so much the better for the hypothesis if we do.
(c) If the evidence is not itself very likely, then it is significant when it does turn up. Surprising evidence is good evidence.

It is perhaps best to see (b) and (c) together – that is, as the ratio of $p(e/h)$ to $p(e)$. This ratio is the measure of how much more likely the evidence is given the hypothesis than it is independent of the hypothesis. If this ratio is around 1, then $p_{after}(h) = p_{before}(h)$, i.e. no change is the degree of belief. This is the case where the hypothesis gives us no special reason to expect the evidence or to think it unlikely. On the other hand, if $p(e/h) > p(e)$, then $p_{after}(h) > p_{before}(h)$. We increase our degree of belief because the evidence we obtained is more likely given the hypothesis than we would otherwise have thought. The third case is $p(e/h) < p(e)$, so that $p_{after}(h) < p_{before}(h)$. This is the case of negative evidence. Here the evidence is supposed, according to the hypothesis, to be more unlikely than we thought on the basis of background knowledge alone. But we obtained it nonetheless, and accordingly we reduce the strength of our belief in the hypothesis. Imagine I have a patient who has three symptoms: very low pulse, dilated pupils, and chest pains. I suspect a certain virus. People infected with this virus very often have dilated pupils, more frequently than the average patient. So this evidence favours my hypothesis. On the other hand, we would expect someone with the virus to have a very high pulse. So the occurrence of the low pulse, to which we would attach a low probability conditional on the hypothesis, is contrary evidence. As regards the chest pains, there is no reason to expect these in an infected subject. But nor is infection any reason to think chest pains would not occur. So the chest pains are neutral as regards the hypothesis.

These very straightforward features of scientific reasoning are neatly accounted for by Bayes' theorem. Note what Bayes' theorem gives us. Bayesian conditionalization tells us how to change our degrees of belief in the light of new evidence. If conditionalization is *a priori* then we have an *a priori* rule of inductive inference. Not surprisingly Bayesianism has many enthusiastic supporters. However, it also has its detractors. We do not have the space to look at all the criticisms of Bayesianism in depth, but a flavour of the objections is given by the following considerations.

(a) Conditionalization could not be *a priori* because it tells us that a degree of belief at one time (after the experiment) should depend on degree of belief at another time (before the experiment). But *a priori* rules cannot be *diachronic* (across time), they must be *synchronic* (concerning the same time).

(b) The probability we attach to a universal generalization should be no more than the product of the probabilities of the infinitely many independent propositions it entails. As most of these have a probability less than 1, this product is zero. Because $p(h)$ is zero, $p(h/e)$ will also be zero; hence Bayes' theorem is useless.

(c) Bayesianism does not tell us what probabilities to attach to $p(h)$ and $p(e)$. So, even with the same evidence, two researchers could have widely differing degrees of belief in a theory.

(d) As Bayesianism is subjectivist it cannot tell us whether a theory that is well supported by the evidence is thereby more likely to be true. It only tells us that it is more rational to believe it to be true. This, as we have seen with Strawson, is insufficient to resolve Hume's problem.

(e) Scientists may sometimes use Bayes' theorem, but on the whole they do not. Indeed, for many theories they would be hard pressed to put any value on the priors $p(h)$ and $p(e)$, and even $p(e/h)$.

Bayesians like to think they have good answers to these criticisms. Again we do not have the space to look at all of them in detail. But let us start with one where the Bayesian certainly has the upper hand – criticism (b) which comes from Popper. Popper argues that the prior probability of a theory being correct is equal to the product of the probabilities of the individual, independent basic statements it entails. In general, an infinite number of these will remain untested. Each of these must have a prior probability of being true of less than 1, for otherwise we could be certain of their truth in advance of testing. But as these probabilities are less than 1 their infinite product is equal to zero, and so the prior probability of any generalization being true is zero.

The response to this is to point out that Popper has misapplied the relevant rule of probability. He is using a generalized version of the following truth: if C entails both A and B, and the probabilities of A and B are independent, then $p(\mathrm{C}) < p(\mathrm{A}) \times p(\mathrm{B})$. The important condition at work here is that the probabilities of A and B are independent – the truth of one has no bearing on the truth of the other. A might be "I had jam for tea" and B might be "Aldehydes are oxidized to acids by Tollen's reagent". Those two propositions are clearly unrelated. But the propositions entailed by a general hypothesis may very well be related. The hypothesis that bats have large ears entails the two propositions that Horace has large ears and that Hildegard, his sister, has large ears. It would be foolish to assume that these are probabilistically independent. There are many conceivable reasons why the fact that Horace has large ears might be relevant to the probability of Hildegard having the same feature. For instance, one might think it plausible that siblings share many anatomical characteristics. This is a general claim and Popper might argue that its probability of being true is therefore equal to zero, and so cannot be used to establish a probabilistic connection between Horace and Hildegard. But as Popper's argument is intended to establish the conclusion that general propositions have zero priors, it is he who cannot make use of that claim in his argument. It seems that, by "independent", Popper understands not *probabilistically independent* but *logically independent*. It is true that "Horace has large ears" and "Horace's sister Hildegard has large ears" are logically independent. But the logical independence of A and B does not entitle one to use the inequality $p(\mathrm{C}) < p(\mathrm{A}) \times p(\mathrm{B})$ (where C entails both A and B). To gain that entitlement Popper would have to show that probabilistic independence and logical independence are the same thing. But, as the example shows, to do that would require *assuming* that general propositions have zero prior probabilities.

Let us now look at objection (c). The Bayesian response is taken by many Bayesians to show the power of Bayesianism. But I think this also shows up some weaknesses of Bayesianism. The objection is this. Say you and I both agree that a piece of possible evidence e has a certain probability assuming the hypothesis h which we are investigating. We agree that $p(e/h)$ is 0.8. But you think that the hypothesis is as likely as not ($p(h) = 0.5$), while I am more sceptical ($p(h) = 0.3$). You think the evidence is mildly surprising ($p(e) = 0.4$), while I think it mundane ($p(h) = 0.6$). What do we now think about the probability of the hypothesis? We both think it is more likely, but you attach a probability of 0.875 to it, while my adjusted degree of belief is 0.35. So after consideration of the same piece of evidence, you think the hypothesis really quite probable while I am still sceptical. This is because the prior probabilities we each attach to h are different. But as Bayes' theorem does not tell us how to fix our priors it cannot be used to say which hypothesis a person should adopt given the evidence.

The Bayesian has the following response. As the number of tests increases, opinion will converge – and will converge on the correct hypothesis. Let us look at an illustration. You, Lucy, and I are investigating a coin that we think might be biased. We can describe the bias by attaching a probability to it falling heads up. For an unbiased coin this probability is 0.5. We have different initial views on whether it is biased. You think it is biased towards tails, Lucy thinks it is unbiased, and I think it is biased towards heads. The various hypotheses that we consider are:

h_0 the probability of heads is less than 0.2;
h_1 the probability of heads is less than 0.4 but greater than 0.2;
h_2 the probability of heads is less than 0.6 but greater than 0.4;
h_3 the probability of heads is less than 0.8 but greater than 0.6; and
h_4 the probability of heads is greater than 0.8.

The prior probabilities which we attach to these various hypotheses, in accordance with our views about the bias, are:

	h_0	h_1	h_2	h_3	h_4
You	0.1	0.7	0.1	0.08	0.02
Lucy	0.05	0.1	0.7	0.1	0.05
Me	0.02	0.08	0.1	0.7	0.1

So we carry out our first test, which consists of ten tosses of the coin. It turns out that four tosses yield heads. Let us see how we now view the probabilities of the hypotheses, having conditionalized them in accordance with Bayes' theorem:

	h_0	h_1	h_2	h_3	h_4
You	0.007	0.851	0.124	0.018	0
Lucy	0.003	0.119	0.855	0.022	0
Me	0.004	0.256	0.328	0.412	0

We continue our coin tossing until we have thrown the coin 50 times with 18 heads, and then further for a total of 100 throws of which 34 are heads. Our conditionalized probabilities after each test are:

After 50 throws (18 heads, 32 tails):

	h_0	h_1	h_2	h_3	h_4
You	0	0.971	0.029	0	0
Lucy	0	0.401	0.592	0	0
Me	0	0.794	0.206	0	0

After 100 throws (34 heads, 66 tails):

	h_0	h_1	h_2	h_3	h_4
You	0	0.999	0.001	0	0
Lucy	0	0.948	0.052	0	0
Me	0	0.990	0.001	0	0

What these tables show is that we have all converged on hypothesis h_1, despite having started off with very divergent prior probabilities. The idea is that in time our prior probabilities get *swamped* by the weight of evidence. They no longer have much of an influence on what we think. (What this example also shows is that the priors for really bad hypotheses get swamped more quickly. I rapidly came over to your opinion, more quickly than Lucy did, although she started with an opinion closer to yours than I did. In the face of adverse data, the high prior I attached to my favoured hypothesis did no good. And, as I attached similar priors to h_1 and h_2, I was swayed by the evidence in favour of the former.)

This is only an example – the evidence might have been rather different. But it is a highly representative example of coin tossing. Given the fact that the coin is biased in the way it is, the ratio of heads is of the sort that we would most often get. And thanks to the law of large numbers, the more often we toss the coin, the more likely it is that the ratio approaches 0.3.

This is a very impressive result for Bayesianism. It seems not to matter what our priors are. We will in time end up agreeing about what the best hypothesis is; and, as this example shows, that will most often be the correct hypothesis. Nonetheless, it seems to me that Bayesianism is too good to be the whole story.

First, the convergence result we have just seen is not foolproof. If I am so inclined I may choose whatever priors I wish – and if I am extreme enough, it could remain the case that even after lots of adverse evidence my favoured hypothesis still has the highest probability. For instance, had Lucy attached a prior probability of 0.996 to h_2 and 0.001 to each of the other hypotheses, even after the 100-toss trial her conditionalized belief in h_2 gives it a probability of 0.9 while the probability of h_1 has risen only to 0.1. Bayesianism does not tell us what priors to attach, so it cannot rule out extreme priors that prevent rapid convergence. In practice, scientists may not differ so widely in their priors (although in some disputes they might), and we would regard it as absurd to adopt an unusually high or low prior without good reason. But this just goes to show that there is more to the actual practice of scientific reasoning than is accounted for by Bayesianism. Something other than Bayes' theorem must contribute to constraining our priors. In some cases the constraint will come from a previous application of Bayes' theorem. Clearly, however, this cannot always be the case, on pain of infinite regress. It cannot be that every use of Bayes' theorem requires an earlier use of Bayes' theorem.

The coin-tossing example hides much of the complexity involved in the evaluation of scientific theories. We were testing a coin for bias, which we know must be a number between 0 and 1. So all the possibilities are covered by the five hypotheses we considered. In testing a scientific theory things will not be so straightforward. Let us think about the postulation of the neutrino. Pauli first proposed the neutrino to account for discrepancies in the observed energies of β-emission. Fermi worked on the idea and showed that it could accurately account for the observed distributions in energy. Before he did so, Niels Bohr even suggested that the law of conservation of energy did not hold at the micro level, but only for collections of particles. This thought was regarded as highly unpalatable by most physicists. Imagine that we are in the position of someone having to assess the neutrino hypothesis in the light of Fermi's results. What prior probability do we attach to the hypothesis? Our answer will depend on our judgement of several factors. How likely is the existence of an unobserved particle? How successful have postulations of similar particles been hitherto? How well established are any theories with which this hypothesis might conflict? How much would such theories need to be altered to accommodate this hypothesis? How good are alternative explanations – such as dropping the law of conservation? How likely is it that we have thought of all the possible explanations? Do we know this field well enough to estimate the chances of some as yet not thought of alternative hypothesis successfully explaining the same data? Similar considerations will apply to estimating the prior probability of the evidence itself. Furthermore, they will also apply to putting a figure on the probability of the evidence conditional on the hypothesis. Where the hypothesis is a statistical one, such as in the coin-tossing case, this is a matter of straightforward calculation. But this need not always be the case. The meteorite hypothesis seeks to explain the extinction of the dinosaurs. How likely is extinction given a meteorite impact? Difficult to say, as we do not have much idea what would happen, even whether the effect would be overheating (greenhouse effect) or cooling (shutting out the Sun); but it is a reasonable inference that the effects would be pretty drastic either way.

What this is supposed to show is that a lot of hard thinking needs to go into estimating our priors. This hard thinking is (a) inductive and (b) requires the comparative assessment of competing hypotheses – how likely they are compared with the hypothesis being studied, and how likely the evidence is on the assumption of each hypothesis in turn, including a catch-all hypothesis to cover explanations we have not thought of. I am not suggesting that Bayes' theorem is in error, but that its application fails to encapsulate all the reasoning involved in a scientific inference. The use of the theorem demands distinct processes of reasoning and inference in fixing our priors. Furthermore, if Bayes' theorem does not capture all our inductive practices, then neither does it justify them.

That Bayes' theorem does not capture all inductive reasoning can be seen by comparing it with inference to the best explanation. According to the latter, the features that dispose us to favour a hypothesis include simplicity and explanatory unity (see Chapter 2). Bayesianism does not do justice to this. Say we have two pieces of evidence, *e* and *f*, and a hypothesis *h*. Bayesians are concerned only with the probabilistic relations between *h*, *e*, and *f*, and disregard the internal structure of *h*. It makes no difference to Bayes' theorem that *h* should be a unified structure or that it might be equivalent to (*i* & *j*) where *i* and *j* are distinct hypotheses (such that *i* explains *e*, and *j* explains *f*). This disregards our intuitions that a hypothesis which provides a unified explanation of distinct pieces of evidence thereby has some advantage over alternative hypotheses that do not provide a single explanation.

Some Bayesians reject the criticism that Bayesianism does not acknowledge the virtue of simplicity by arguing that the notion of simplicity is too vague and means different things to different people. For this reason it cannot be an important part of scientific reasoning. No doubt they would make similar remarks about the concept of explanatory unity. Nonetheless, it is the case that scientists do respect explanatory virtues such as simplicity and unity. That there are differences from case to case in how these notions are applied is no surprise. I have argued that the notion of "similarity" is one which, as we use it, is relative to our favoured theories. Although things are objectively similar or dissimilar, our knowing this will depend on our success in discovering natural kinds, which in turn depends on the success of our current theories. It will also depend on our explanatory interests, which will determine which theory we regard as significant to a particular problem. So it is no surprise that there is no uniform appreciation of whether two things A and B are similar. For the same reason our notion of simplicity is such that judgements of simplicity will be relative to the theories we are employing.

In fairness to Bayesians it must be made clear that they do not hide the fact that they have little to say about the fixing of priors. On the one hand they may deny that the considerations which lead us to assign priors as we do are relevant to questions of scientific reasoning. All the latter should be concerned with is how we should adjust our view of a hypothesis in the light of evidence. Whether one rejects or accepts this depends on whether one agrees or disagrees that structural features of a hypothesis, such as simplicity, explanatory unity, and integration with other hypotheses, can be part of the rational evaluation of a hypothesis. I have argued that they should be, and thus that Bayesianism is incomplete. On the other hand, Bayesians may accept that Bayesian conditionalization is open to supplementation – for instance by a theory of how structural features should contribute to the evaluation of the prior probability of a hypothesis. In contrast, Inference to the Best Explanation as I have described it does give a role to the structural features left unaccounted for by Bayesianism, and so in that regard at least provides a better understanding of the rationality of scientific inference.

Lastly, it should not be forgotten that, as a theory of subjective probability, Bayesianism has little to say about the propensity of inductive practices to lead to the truth. We may well agree that someone who adjusts the credence they give to a hypothesis in accordance with Bayesian conditionalization is thereby behaving rationally. But nothing in Bayesianism can show that such a person is also more likely to end up with true hypotheses than someone who infers differently. In this way, whatever the merits of Bayesianism as either an explanatory description of our inductive practices or an account of inductive rationality, it is in the same boat as Strawson's account when it comes to finding an answer to Hume's problem. Neither shows whether rationality has a propensity for success. Consequently, neither tells us whether inductive knowledge is possible. That is the task of the next chapter.

Further reading

Hugh Mellor, in *The matter of chance* presents a personalist view of probability combined with a theory of objective chance. Classical probability and statistics are explained in Richard von Mises' *Probability, statistics and truth.* The Bayesian view, along with criticisms of classical probability, is expounded by Colin Howson and Peter Urbach in *Scientific reasoning: a Bayesian approach*. John Earman's *Bayes or bust* is a very thorough but quite difficult examination of Bayesianism and the alternatives. Clark Glymour's "Why I am not a Bayesian" tells us just that.

Chapter 7

Inductive knowledge

In the previous two chapters we have looked at attempts to solve or avoid Hume's sceptical problem of induction. The time has come to show where I think our best hope of a solution lies. First, we need to be clear about what the problem is. Sometimes the problem is stated as that of "justifying induction". I think this is potentially misleading. There is no one thing "induction" which stands in need of justification. Strawson was right to point out that there are many inductive arguments, some of which are reasonable and others which are not. We also saw that any attempt to construct a super-inductive argument, for instance one using the uniformity premise, is doomed. Any such super-inductive argument is a bad inductive argument, as it clearly leads to false conclusions. Secondly, if we mean by "justification" the task of showing that an inductive argument will lead to or has a propensity to lead to true conclusions, then that task is not a philosophical one. Even if it is the case that we can have *a priori* grounds for thinking that an inductive argument is reasonable (as Strawson claims), we cannot have such grounds for thinking that someone who uses that argument will thereby achieve a high proportion of true conclusions. For that is a contingent issue. The world *might* be irregular. There *may* be no laws of nature. Hence any inductive argument might be utterly unreliable. Therefore, no philosophical argument can show that it is reliable. If the proposition that *p* is a contingent proposition of science, then the philosophical problem is not: *Do we know that p*? Rather it is: *Could we know that p*? That is to say, this issue is one of showing that circumstances could exist in which we do have knowledge as a result of using an inductive argument. In addition, of course, we would like to know whether such a possibility is consistent with what we do know of the circumstances we actually find ourselves in.

So the philosophical problem is that of showing the possibility of knowledge. Hume's problem suggests that inductive knowledge is impossible. The inductive sceptic requires that, in order for an inductive argument to yield knowledge, I must be able to demonstrate that my use of induction will yield knowledge. This demonstration must itself amount to knowledge. So, in order for induction to yield knowledge, I must know that it yields knowledge.

Later on in this chapter I will outline a view of knowledge according to which this is not a requirement on knowledge. I will claim that one can use a method to obtain knowledge without also knowing that the method is reliable. For the moment I want to relate what has just passed to a principle that is often mentioned in epistemology and I think is clearly false. As we have seen, the sceptic wants us to be able to show how we know what we claim to know. Thus, if we know something, we know that the method we used is one which gives us knowledge. If we reflect on what we believe and how we came to believe it, then if that belief amounts to knowledge we will also know that this belief is knowledge.

The requirement on knowledge, that if someone knows something, then they know that they know it, is called the K-K principle. If we denote "N knows that *p*" by "K*p*", then the principle can be formalized thus: $Kp \rightarrow KKp$. The foregoing remarks are intended to suggest that inductive scepticism (and indeed other forms of scepticism) imply the K-K principle or something like it. As I have presented the matter, the sceptic's position implies a slightly weaker version of the K-K principle, i.e. that if I know something, then I know enough to know that I know. I think that the K-K principle, in either form, is false.

For instance, does Lucasta know that hydrogen, helium, lithium, and so on, are the first 50 elements? I might find out by asking her to write down a list of the first 50 elements and then checking her list against the periodic table. If the list is right I now know that she did know. So we have two cases of knowledge: her knowing that these are the elements and my knowing that she knows them. Clearly we could have the former without the latter – presumably she knew the elements just before writing them down, but I did not then know that she knew them. What is true for me may also, I contend, be true for Lucasta. Before writing down the list she may not know whether she knows them. For instance, she may think "I learned the periodic table at school, but that was a while ago and it is quite possible that I have forgotten it". On testing her memory she discovers, just as I did, that she did indeed know the list of elements. So, before the test, she knew them but did not know that she knew them. Nor did she know enough to infer that she knew them. She did know the list of elements, but had to wait for the evidence of the test before coming to know that she knew them. If this is a fair description of such a case, then we have a counter-example to the K-K principle.

Another case might be this. I use an ammeter to measure the current passing through an electrolyte. It gives a reading of 0.2 A. Do I know the current is 0.2 A? What does that depend on? It depends on the ammeter being well designed, on its not being defective, and so on. If the ammeter is in good working order, then I know that the current is 0.2 A. Do I know that I know what the current is? Probably not, if I have not checked the ammeter today and if I know nothing about its design or prior reliability. This is the way we use the word "know". We do not regard it as necessary to study the theory of ammeter design or to make checks on reliability before we can use an ammeter to know what the current through the solution is.

Nonetheless, the sceptic wants to deny that in such a case I really do know that the current is 0.2 A. The sceptic will say that I do not have this knowledge unless I have made checks on the reliability of the ammeter. Furthermore, I will need to know that the method I used to check the reliability of the ammeter is itself reliable. And so on. Clearly the sceptic's demands cannot be met. But there is no reason why we should try to meet the sceptic's demands. As the counter-examples to the K-K principle show, we do not use the word "know" in the way the sceptic expects or requires. It is not that our use is sloppy or mistaken. Rather it simply reflects what the word "know" means. For this reason, rather than struggling to meet the sceptic's impossible conditions, it is a far more fruitful approach to see whether there is a plausible account of the meaning of "know" that accords with our normal usage, rejects the K-K principle, and therefore can avoid inductive scepticism.

In this chapter I will present a view of knowledge which seeks to do this. This view is known as *reliabilism.* I myself think that reliabilism is imperfect. Nonetheless, it is instructive to see how a plausible epistemology can overcome Hume's problem and provide satisfying answers to other problems in the epistemology of science.

Reliabilist epistemology

Before seeing what reliabilism says, I want first to prepare the ground by providing a motivation for its claims. We start with the concepts of *truth* and *belief.* Intentional action is typically explained by referring to a combination of someone's goals, aims, or desires and their beliefs. So for instance, Darren's drinking port is explained by his desire to get drunk and his belief that drinking a bottle of port will inebriate him. Truth is important here because truth makes the difference between achieving one's goals and the possibility of failure. As Darren's belief is true, then his action, if properly carried out, will ensure that his desire is met. The truth of a belief is sufficient to secure the success of an action in achieving the relevant desire. Falsity, however, leaves open the possibility of failure. If Darren mistakenly thought that mango juice would get him drunk, then he is liable to fail in his aim.

Of course, Darren might still achieve his aim despite a false belief, by good luck. For instance, someone might have laced the mango juice with vodka. So falsity does not guarantee failure. Nonetheless, truth does guarantee success. (A proviso needs to be made here. Often we act on probabilistic beliefs. I am late for a train. I think that catching a cab rather than a bus to the station will maximize my chances of catching the train. The truth of this does not ensure that my desire to catch the train is satisfied. Success is not ensured. But if the belief is true then the chances of satisfying my desire will be maximized.)

True belief therefore looks very desirable. Note, however, that true belief does not constitute knowledge. For instance, on drinking the laced mango juice, Darren believed truly that he would get drunk. But he did not *know* that he would. Accidentally true belief is not good enough for knowledge. But if true belief is good enough for success, why want anything more? Let us see why. Consider the next morning. Darren is hungover. He desires to be rid of his hangover. However, unlike the night before he has no beliefs relevant to his desire – he has no beliefs about how to cure hangovers.

So Darren has a new desire, the desire to acquire a new true belief, about how to cure a hangover. As we saw a moment ago, to achieve his desire he needs a corresponding true belief. This will be a true belief about how to acquire true beliefs. So, for instance, Darren may believe that reading a medical encylopaedia is a good way to get true beliefs about hangover cures. As long as Darren is right about this, that his medical encylopaedia is reliable, then reading it will satisfy his desire of getting true beliefs.

Let us review this chain of desires and beliefs:

(a) N has desire *d*;
(b) N has belief *b* that doing *a* will achieve desire *d*; and
(c) if belief *b* is true then N's doing *a* will achieve *d*.

Our first case was where *d* was the desire to become inebriated, *a* was drinking port, and *b* was the belief that drinking port will cause inebriation. The second case was where *d* was the desire to have a true belief about hangover cures, *a* was reading a medical encyclopaedia, and *b* was the belief that reading the encyclopaedia would give a true belief about hangover cures. In both cases (c) says that the truth of the relevant belief is sufficient for the desire in question to be satisfied. Note that it does not say that knowledge is required. In the second case, in order to get true beliefs about hangovers it is sufficient that it be true that the encylopaedia is reliable. Darren does not have to know that it is reliable.

Reading the encyclopaedia is Darren's method of acquiring medical beliefs. He wants the encyclopaedia to be reliable. If in general he wants to acquire true beliefs, it is no good using an encyclopaedia which is full of falsehoods.

This is where knowledge comes in according to *reliabilist* epistemology. Knowledge is not just true belief, but true belief *acquired by a reliable method.* This neatly explains the role of the concept of knowledge: we want to acquire true beliefs (for instance, in order to satisfy desires); knowledge is what we get if we use a reliable method to get those beliefs. We can also see why knowledge is preferable to mere true belief. Consider Darren's belief that the glass of mango juice would inebriate him, which is true because the juice has secretly been laced. Darren's belief rests upon a false general belief that mango juice is intoxicating – and so his belief about the glass of juice does not come from a reliable method, which is why it is not knowledge. Now reflect upon the fact that Darren might easily discover that mango juice is not intoxicating – he might read this on the carton, or in a book, or be told so by someone. Once he acquires the true general belief that mango juice is not intoxicating, he will abandon the (fortuitously true) belief he has that this glass of juice will make him drunk – indeed he might acquire the false belief that that glass will not make him drunk. The point of the example is that an accidentally true belief is in danger of being overthrown by evidence which shows that it originated in an unreliable fashion. Reliably produced belief – knowledge – is not so easily susceptible to evidence that might make us abandon it.[49]

It is important to note that if I get knowledge because I acquire my true beliefs by a reliable method *m* (e.g. an encyclopaedia), it is required only that *m* be reliable as a matter of fact. It is not required that I know that the method is reliable – for instance Darren does not need to know that the encylopaedia is reliable. A reliable method will yield true beliefs whether or not I know that it will – and true beliefs are what I want. That is, I can know without knowing that I know. The K-K principle is rejected.

We can summarize what we have learned about knowledge thus:

(K) N knows that *p* if and only if

N believes that *p*;
it is true that *p*; and
N's belief that *p* was acquired by a reliable method.

Reasoning with induction

Now consider the case where the belief that *p* is the conclusion of an inductive argument: Darren reasons that he will suffer from a hangover because he has always had hangovers after drinking port, that people often have hangovers after drinking liberal quantities of alcohol, and because he has just consumed a bottle of Warre's 1963 vintage port. Now for this belief to count as knowledge, it must be true and reliably acquired. And so, if reasoning inductively in this way is a reliable method of acquiring beliefs, and if the belief is true, then his belief counts, by (K) as knowledge.

Let us return to the initial question about inductive knowledge. Is it possible? According to the analysis (K) of knowledge, we may answer yes. If inductive arguments of this sort are reliable, and if Darren's belief is true, then Darren has knowledge. So for it to be possible for Darren to know something inductively, it is necessary that it be possible both that inductive arguments of this sort are reliable and that his belief is true. Now it seems clear that it is possible that such inductive reasoning is reliable. For instance, if it is a law-like truth that people with Darren's constitution will have hangovers after drinking so much port, it will be reliable to argue as he has. If that is the case, then by (K) he will know that he will have a hangover. This shows that it is possible, if the circumstances are right, for someone to have inductively gained knowledge. Remember that our philosophical task is to show that inductive knowledge is *possible* – it is a different, non-philosophical question whether or not we do actually have inductive knowledge. But we can say that the possibility of knowledge is not obviously a mere unactualized possibility. From what we can tell of the world, it might be one in which there are laws of nature governing alcohol and hangovers, and it might be one in which Darren's reasoning is reliable. So he might well have inductive knowledge – a conclusion which contradicts the sceptic's claim that he cannot possibly have knowledge.

According to the reliabilist the sceptic's error is the claim that, if I know something, I must know that the process by which I acquired this knowledge is reliable. For instance, the inductive sceptic required that, in order to know something using induction, I must be able to say why induction is likely to lead to the truth. On the reliabilist account, the method by which I acquire my beliefs must be reliable, in order that they should count as knowledge; but it is not necessary that I should know that it is reliable. Reliabilism is an *externalist* view about knowledge and justification. Whether or not the method I have used is reliable will depend on factors outside my mind, and that is why it is something of which I may be ignorant.

Observation and theory

Not only does reliabilism allow us to avoid Humean scepticism about induction, but other sceptical worries can now be seen to be misplaced. Recall the problem of cognitive dependence of theory on observation, the theory ladenness of observation as it is sometimes called. This gave rise to sceptical worries. If observation depends on theory and knowing the truth of a theory depends on observation, where does the process begin? How can we get off the ground? For instance if one starts by trying to observe, some theory will be assumed. But because this is the start, one will not know that theory. And if one depends upon it in observation, how can that yield knowledge? And if observation cannot deliver knowledge, how can we get to know any theory for which that observation is supposed to be evidence?

Armed with reliabilism we can see where this reasoning fails. It assumes that for a theory-laden observation to yield knowledge the observer must know the theory that is being used to be true. According to reliabilism this is not necessary. To get observational knowledge it is sufficient that the observational method be reliable. This may require that the theory embedded in the observation be true or almost true, but it will not require that the observer know this. For instance, one might use a radiotelescope to make astronomical observations or an electron microscope to look at microbes. Do the observers need to be experts on the principles on the basis of which these instruments were designed? Not necessarily. If those principles are true and other aspects of the design satisfactory, then the instrument will be reliable. That is all that is required for observational knowledge.

One way to see this is to consider that the theory in question need not always be true. It can be false, and *a fortiori* not known to be true. Many instruments have been designed employing Newtonian mechanics or classical electrodynamics, which are, strictly speaking, false. But that does not worry a reliabilist. For a false theory can be used in the design of a reliable instrument so long as the theory is sufficiently close to the truth that the results generated by the instrument are not significantly affected.

Often the theory or assumptions with which a theory is laden will be tacit – the observer will not be consciously committed to those assumptions. In some cases our expectations are innate. In other cases they will be learned, but so commonplace to the observer that he no longer thinks about them. A young microbiologist may at first have consciously to think of what he has learned about spirochetes in order to recognize them under a microscope. But as he becomes more experienced the necessity of recalling this theory to mind disappears – recognition is instantaneous. In yet other cases, as I have mentioned, the theory on which the observation depends may be known only by the designer of the equipment being used. In these cases it may difficult to identify precisely what our theoretical commitments are, let alone whether they amount to knowledge. Reliabilism tells us that this does not matter. What matters for an observational method to give us observational knowledge is whether the method is reliable. We do not need to be able to say precisely what method we are using, nor do we need to be able to say what its theoretical assumptions are, nor do these commitments themselves need to amount to knowledge.

Innate epistemic capacities and reasoning about induction

This suggests that reliabilist epistemology can also help us with the problem with which we concluded Chapter 3. In outline the problem was this. We know what natural kinds there are by seeing which properties appear in the laws of nature. But one lesson of Goodman's problem was that we cannot identify the laws of nature without some prior identification of natural kinds. (Different arbitrary choices of natural kinds would lead to mutually inconsistent sets of natural laws.) We were faced with another chicken-and-egg problem. The solution I suggested was that we have innate dispositions to see certain things as similar and to expect certain situations to repeat themselves. We are able to detect certain natural kinds and laws, which we then use as the basis for discovering less apparent laws and kinds.

A worry that might arise is this: What if our innate dispositions are faulty? Might we not be erecting the edifice of science on shaky foundations? Do we not need to know that our dispositions are reliable in order for them to provide a basis for scientific knowledge? Armed with reliabilism we can see that we can reply negatively to the last question. A reliable method can produce knowledge, even if we do not know it is reliable. The fact that our dispositions *might* be unreliable does not mean that we do not *actually* have knowledge.

This takes us some way towards a resolution of Goodman's problem. One lesson to draw from the problem is that there is no model of induction. There is no template that allows us to say whether a proposed inductive argument is a reasonable inductive argument. This fits with our conclusion that there are no models of explanation, confirmation, or lawhood. The best we can do as regards a description of induction is inference to the best explanation and, as I have been emphasizing, that is a holistic enterprise.

Thus we cannot tell just from the structure of the two arguments that: (a) all observed emeralds are green, and therefore all emeralds are green; and (b) all observed emeralds are grue, and therefore all emeralds are grue, which deserves our credence. It will depend on which fits in best with the optimal systematization of everything we know in a structure of laws, explanations, and natural kinds. It is clear which of (a) and (b) does fit in best, as a matter of actual fact. The property green already plays a role in our theoretical structures, while grue does not; in particular those structures justify our innate capacity to detect and distinguish green things.

It is therefore a scientific question, not a philosophical one, which of the two arguments it is reasonable to adopt. This may appear circular or question begging, since the existing scientific framework is built upon green-like (and non-grue-like) inferences. Say we thought we had some reason that made us unable to decide between the green inference and the grue inference. Then that would be a reason not to put our faith in the existing framework of science, which is based on green-like inferences. This objection is to return to a more general, Hume-like scepticism. It suggests that if science or our intuitions are to justify one inference or the other, then we have to be able to justify that science or those intuitions. The requirement that to be justified one must be able to justify one's justification is similar to the K-K principle which we rejected above – it is a J-J principle, which I shall mention again below. From the point of view of reliabilist epistemology, it is sufficient that our science be reliable for it to justify the green inference over the grue inference.

The key to reliabilist solutions of the problems of epistemology is its externalism. That is, whether a hypothesis counts as knowledge (or an inference counts as justified) depends on the actual reliability of the methods that support our belief in the hypothesis or the actual reliability of inferences of that sort. That said, I think we can say more. Among the things we can investigate are our own methods, belief-forming processes, and epistemic dispositions. We can come to know whether our methods are reliable – if the methods we use in our investigation are themselves reliable. There are different ways of coming to see that a method is reliable. Most straightforwardly, we can just keep using the method, checking to see that it gives accurate results. With more sophistication one can investigate the mechanism underlying the method in question. Knowledge of this functioning may allow us to judge whether or not it is reliable. For instance, faced with an instrument of unknown reliability, one can simply test it to see whether it does give the right results. Or one can take it apart to see whether it is properly designed and constructed for the job in question.

With regard to our basic epistemic dispositions, we have both ways of investigating their reliability. One can assess how reliable one's eyesight is or whether one has perfect pitch just by trying to read signs at different distances or trying to name pitches played on a piano. More fundamentally, physiologists can explain the general reliability of perceptual capacities. Thus science can give theoretical justification to its starting point. We can say that we know that our perceptual faculties are, by and large, reliable, and that they do pick out certain natural kinds and natural laws. The appearance of circularity here is not a cause for concern. This is because the theoretical understanding of perception is not necessary for perception to be a solid foundation for knowledge – for it is sufficient that perception be reliable. If perception is reliable and if the other methods of science that are built upon perception are themselves reliable, then physiology in particular is reliable. In which case we can use it to know that perception gives us knowledge.

The circularity of science investigating itself is non-vicious, because the circle is not necessary for knowledge; indeed, it has a positive epistemic effect. For instance, someone might wonder whether they have lost a contact lens (with which their eyesight is reliable, but without which it is not). It is perfectly reasonable for such a person to look in the mirror to see whether the lens is in place. If it is in place, they can know by looking that it is. One way the scientific investigation of both our innate capacities and the methods of science can have positive epistemic effect is by weeding out unreliable methods and improving and extending others. More will be said about this in the next chapter, which looks at progress in science.

Problems with reliabilism

I think that reliabilism shows us a way out of the epistemological problems of science. However, more needs to be said about reliabilism. In the view of some philosophers there are objections to which reliabilism does not have a satisfactory answer. If knowledge is a matter of a belief acquired by a reliable method, what then do we mean here by "method" and when is a method regarded as "reliable"?

Let us look at the second of these two questions first. The short answer to the question "When is a method reliable?" is "A method is reliable when it tends to produce true beliefs". The presence of the word "tends" suggests that a method is reliable, even if it sometimes produces false beliefs. The rationale for this approach is that few, if any, of our belief-producing methods are 100 per cent reliable. Sometimes we make mistakes with our unaided sense. But we would not want to say that the possibility of visual error means that one's eyesight is an unreliable mechanism. For, if one said that, then one would have to conclude, according to reliabilism, that one never had perceptual knowledge. So we have to allow for the possibility that a method might sometimes go wrong. But it would still have to be right in a sufficiently high proportion of cases. Someone who more often that not makes mistakes about the way things look clearly has unreliable eyesight and would not be said to be able to see things.

So we can use a method to give us knowledge, even if the method might in a small proportion of cases go wrong. But now a conundrum arises. Say we use a method which is reliable but imperfectly so. For instance an experienced radiographer may be using magnetic resonance imaging to detect cancerous tumours. Let us say that the combination of the technique and the radiographer's skill means that their accuracy is high. The nature of the cancerous brain tumours is such that they cause a characteristic image when scanned, which the radiographer has much experience in detecting and distinguishing from images of different lesions or tumours. But on very rare occasions the radiographer misidentifies a non-cancerous tumour in a patient's brain as being cancerous. Consider just such an occasion. A benign tumour is identified as cancerous. The radiographer has the belief "This patient has cancer." In this case the tumour was non-cancerous. But, in another part of the body, the pancreas, which has not been examined, there is cancer in its early stages. So it is true that the patient has cancer. What we have then is a belief, that the patient has cancer, which is brought about by a reliable method and which also happens to be true. So according to the reliabilist analysis of knowledge given above, the radiographer knows that the patient has cancer.

Nonetheless, this cannot be right. The cancer in the pancreas has nothing to do with the non-cancerous tumour in the brain and it is the latter which led to the mistaken diagnosis of cancer. The right thing to say is that the true belief in the existence of cancer in the patient is only true by a chance coincidence, the coincidence of a rare misdiagnosis of the brain tumour with a cancer elsewhere. Coincidentally true beliefs like that, even if brought about by a reliable method, cannot count as knowledge.

So the reliabilist analysis needs a repair. Let us reflect on what is needed for reliability. Could there be mere chance reliability, sufficient for knowledge? In Chapter 1, I argued that there could be regularities that give the appearance of being laws, but which in fact are merely accidental, pure chance coincidences. Imagine such a set of coincidences, which someone mistakenly takes for a law – as for instance more than one astronomer took Bode's law to be in the late 1700s.[50] And imagine such a person made predictions using it. As there is a regularity, albeit accidental, the predictions will turn out to be true (as astronomers correctly predicted the existence of "planets" between Mars and Jupiter – the asteroids Ceres and Pallas and the orbit of the next planet beyond Saturn and Uranus). Do such true predictions founded on an accidentally "reliable" method count as knowledge? Did these astronomers know that there were heavenly bodies between Mars and Jupiter? Surely not. Again the truth of the prediction is a lucky coincidence. This contrasts with the calculations that allowed Adams and Leverrier independently to predict the position of an unseen planet (Neptune), which was then observed where predicted. In this case it seems reasonable to say that Leverrier did know where the new planet would be observed. What is the difference? Clearly the difference is that in one case the method used was only apparently or accidentally reliable (the Bode's "law" case), while in the other the method was founded on a genuine set of laws (Newton's laws of motion and gravitation).

So what is meant by the reliability of a method is not just that the method gives the right answer almost all the time, but that this fact is backed by or explicable in terms of the laws of nature. Remember though, that a person does not need to know that the method they are using is reliable for it to give them knowledge. And so a person need not know which laws explain the reliability. One may use a method, find that it is reliable, and then later investigate the natural processes which explain that reliability. For instance, we use our power of hearing, and find that for most of us it is reliable. It is the job of physiologists and cognitive scientists to find out why it is reliable.

The fact that the reliability of a method has to be a law-like, naturally explicable property of the method allows us to see a solution to the problem raised by the radiographer case. In that case the radiographer had a true belief obtained via a reliable method, but we do not regard it as knowledge because the method on this occasion was faulty and the belief turned out to be true only by coincidence. In cases where the method is working properly and the radiographer does know whether the patient has cancer or not, then the success of the method on such occasions can be explained. The explanation of the success of the method on individual occasions will be the same as the explanation of the general reliability of the method. Some property of cancerous tumours plus the functioning of the magnetic resonance scan means that they yield a characteristic image. We can add the fact that the training and experience of the radiographer means that the radiographer's ability to spot these images is law-like too. These facts, which could be fleshed out in more detail, *explain* why this method of detecting tumours is reliable. Let us take an ordinary occasion of a successful detection of a tumour. The explanation on this particular occasion will follow the same pattern – the property of a tumour that gives it a characteristic image, which the radiographer is able to identify.

This contrasts with the accidentally correct diagnosis of cancer on the occasion of a misidentification of a tumour coincidental with an undetected cancer. In this case the the "success" of the diagnosis will have quite a different explanation. That explanation will *not* refer to the features of tumours which yield the tell-tale scan under magnetic resonance imaging, because the actual cancer in this case was not picked up by the scan. Instead the explanation will have two disconnected parts: (a) why it was the non-cancerous lesion caused an image of a sort that the radiographer was able to mistake for a cancer, and (b) mention or explanation of the (independent) fact that the patient has pancreatic cancer.

So the new, improved version of reliabilism has to say that to yield knowledge the method must be successful on the occasion in question and, furthermore, the explanation of its success is in outline the same as the explanation of its reliability in general. So, in short:

(K2) N knows that *p* if and only if

N believes that *p*;
it is true that *p*; and
N's belief that *p* rests on a reliable method such that the explanation of the general reliability of the method also explains its success on this occasion.[51]

I am not convinced that (K2) gives a necessary condition for knowledge, but it does effectively illustrate the sort of position reliabilists adopt. In any case a general problem faces reliabilism, for we have not yet said what counts as a method. Imagine that Paolo believes that alkaline solutions will turn any piece of paper blue. On some particular occasion he dips a piece of paper into a solution and its turns blue. So he believes that the solution is alkaline. On this occasion it just happens that he used a piece of litmus paper, which was why the paper turned blue. If we describe the method as "dipping a piece of paper into a solution, and believing that the solution is alkaline if it turns blue", then his method is unreliable (for instance this method would have him believe that blue ink is alkaline). On the other hand, we could describe the method more narrowly, as "dipping a piece of litmus paper into a solution, etc.", in which case Paolo has used a reliable method. So did Paolo use a reliable method (and hence know the solution is alkaline) or not (and so not know it is alkaline)?

I think that in this case we want to say that Paolo was using the unreliable method since this was the method he took himself to be using. But this may not help with formulating an improvement to (K2). For it is not clear that people can clearly identify the methods they themselves are using. For instance, in sense perception we do not usually reflect on the perceptual processes we use. One can imagine the cross-examination of a witness who says that the accused was at the scene of the crime. When first asked how they knew, they might first say that they looked out of the window and saw the accused. Then it might transpire that the witness was not wearing their spectacles, and thereafter that the accused was standing 20 yards away in broad daylight (and that the witness needs spectacles only for reading). So it was not clear to the witness whether the method they employed was looking, looking without their spectacles, or looking at the accused 20 yards away in broad daylight.

One route to a possible solution is to see that, whatever the method is, there must be an explanation of why the individual possesses and uses that method. This is important because we would not count as knowledge beliefs acquired by a reliable method which was itself chosen only at random. Rather, we want the explanation of the possession and use of the method to have something to do with its reliability. In some cases the explanation will be that we have tried the method and found it to be successful. In other cases it may be that we designed the method using knowledge of the laws that account for its reliability (think of designing a resistance thermometer). For basic perceptual methods of belief acquisition the explanation may have a Darwinian flavour. The reliability of a perceptual organ and a perceptual disposition will explain why these have evolved.

So an answer to the problem of individuating methods might be along these lines. Say it is proposed that N knows that *p* because N used a reliable method *m* in coming to believe that *p*. This will be acceptable if the explanation of N's use of *m* refers to the reliability of *m*. So consider Paolo again. It would be wrong to say that Paolo knows that the solution is alkaline on the grounds that he used a reliable method, the litmus test. For the reliability of the litmus test does not explain why Paolo used it. What explains why Paolo used litmus paper is his false belief that any paper turns blue in an alkaline solution.[52]

Internalism and justification

So far I have couched the reliabilist response to inductive scepticism in terms of knowledge. The reliabilist claims there can be knowledge gained by using inductive methods, and I have been taking the sceptic to deny this. However, the debate is sometimes characterized as a debate about justification, which is the notion Hume uses. Justification and knowledge are clearly different. One can have a false belief that is nonetheless justified (for instance we may allow that Maxwell's belief in the electromagnetic aether was justified, even though we take it to be a false belief). But one cannot have a false belief that counts as knowledge. By reflecting on this, one may take justification to be like knowledge but without the requirement of truth. So looking at (K2), the improved reliabilist definition of knowledge, we can supply a simple definition of justification:

(J) N is justified in believing that *p* if and only if

N believes that *p*, and
N's belief that *p* rests on a reliable method.

This would then allow us to simplify (K2) to (K3) thus:

(K3) N knows that *p* if and only if

N has a justified true belief that *p*, and
the explanation of the general reliability of the method that justifies the belief that *p* also explains its success on this occasion.

The point of the second clause, as in (K2), is to differentiate knowledge from accidentally justified true beliefs.

The notion of justification defined in (J) is an externalist one, like the reliabilist conception of knowledge. If one is justified, one's justification may depend on factors about which one may know or even believe nothing. So long as one's belief is brought about a method which as a matter of fact is reliable, then one's belief is justified. This may be the case even if one does not know that the method is reliable, and even if one is not justified in thinking that the method is reliable. Indeed, one may not have a clear conception of the method at all.

Some philosophers object to this notion of justification. If a person has no clear idea of what method they are using, let alone a good reason for using it, how can it give their beliefs justification?

I think there are at least two reasons why this appears to be worrying. First, it seems as if someone could have justification by accident – they unknowingly adopt a method which happens to be reliable. But, if we adopt the principle (discussed above) that for something to count as a reliable method for the purposes of knowledge and justification the reliability of the method must explain its possession and use, then it could not be entirely by accident that one uses a reliable method to gain justification.

Secondly, I think that some philosophers conflate the justification of the belief and the justification of the method which produces it. In effect I think they adhere to a J-J principle. They hold that for a method to yield justified belief, the method too must be justified, and the reason for holding this is the following intuition. For something to count as a justification, its being a justification should be apparent to the thinker. The justification should not depend on facts outside the thinker's mind. This is *internalism* about justification. At the end of this section I mention my reasons for being doubtful about internalist justification. Nonetheless, I do want to make some remarks that may appease those who think that the notion of justification should be more internalist than is the notion of knowledge.

Here the challenge is to show that there is an *a priori* component to justification. I think that this challenge can be met, for some scientific beliefs anyway. The cost of meeting this challenge is that the link between justification and knowledge is weakened. In this internalist conception of justification, one may have a justified belief, even when the reasoning that produces it is thoroughly unreliable. Consequently, (K3) will not be true (for this conception of justification).

Recall Strawson's answer to the problem of induction. He thought that it was possible to give *a priori* standards of reasonableness for inductive arguments. I argued that Strawson's argument failed as a solution to Hume's problem, as I had presented it, because he could not show that reasonableness is truth tropic (tending towards the truth). Nonetheless, Strawson's argument suggests that *a priori* considerations might tell us whether an inference is *reasonable* given certain evidence, even if those considerations cannot tell us whether the conclusion is likely to be true. This sense of "reasonable" will be equivalent to the new, internalist sense of "justified" introduced in the previous paragraph.

Strawson said that an inductive inference will be reasonable insofar as the evidence strongly supports the conclusion, and unreasonable insofar as the evidence is weak; the strength of evidence is related to the number of favourable instances and the variety of the circumstances in which they have been found. These claims, he says, are part of what "reasonable" means, and so are analytic and *a priori*. I think we can improve on this.

In Chapter 1 I outlined a criterial view of laws. The criterial view adopted some aspects of the Ramsey–Lewis version of the regularity theory of laws. The criterial view accepts that there is a logical relation between certain sorts of regularity and the existence of laws. For the regularity theorist this logical relation is the simple one of identity – laws just are certain sorts of regularity. According to my criterial view, the relation is more complicated. It is part of the meaning of "law" that observed regularities of the appropriate sort are good evidence for the existence of the corresponding laws. I need to emphasize that this is all *part of the meaning*. Someone who understands what "It is a law that Fs are Gs" means, will thereby appreciate that, other things being equal, the existence of Fs that are Gs constitutes some reason to believe that there is a law that Fs are Gs, and that the strength of this reason increases as there are more Fs that are Gs. Also, part of the concept of law is that the strength of this reasoning also increases insofar as we are able to integrate this putative law that Fs are Gs into the overall system of what we take to be laws, and insofar as we are able to do so simply. At the same time the strength of the putative law speaks in its favour. But against it would be alternative explanations of the Gness of the various things that are both F and G (thus obviating the need for the putative law). The details are given in Chapter 1, but I hope this is sufficient to make clear the connection between the criterial view and Strawson's answer. The criterial view adds various elements to Strawson's account, but both agree that the claims just made are *a priori*. They constitute the meaning of the phrase "law of nature".

So the criterial view can accommodate the idea that certain inductive inferences are justified, where justification is construed as an internalist notion. Repeating what I said in Chapter 5, I do not think that this is any solution to the problem of induction, since the fact that an inference is justified in this sense does not make it any more likely to yield a true conclusion (from solid evidence); the answer to problems of that sort rests on externalist arguments like those given earlier in this chapter. Just as I think the K-K principle is flawed, I think the J-J principle is flawed also. For these and other reasons I am not entirely sure that the internalist notion of justification is the one we actually employ. Many philosophers take the notion of justification to be more important or more basic than the concept of knowledge, and think that we should concentrate on the philosophical understanding of scientific justification rather than on scientific knowledge.[53] This notion too may be mistaken. For a start, a plausible canon of rationality is: don't believe something you do not know to be true. Scientists have as their aim not justified beliefs but knowledge. Understanding the world involves knowing things about it – having well-justified theories does not constitute understanding if that justification fails to amount to knowledge. On this view knowledge is what we aim at in having beliefs; our beliefs are justified if their basis is the sort of basis which, in normal circumstances, leads to knowledge.

Further reading

David Papineau, in *Reality and representation*, gives an account of reliabilist epistemology and naturalized epistemology, which are also discussed in Nicholas Everitt and Alec Fisher *Modern epistemology*. An earlier version of these views is to be found in Armstrong's *Belief, truth and knowledge*. Hugh Mellor's paper "The warrant of induction" argues for an inductive justification of (Humean) induction.

Chapter 8

Method and progress

In the previous chapter we looked at the reliabilist claim that knowledge is true belief supported by a reliable method. The mention of method raises the following question: Is there a method distinctive of science – *the* scientific method? The scientific method, it is often claimed, is what distinguishes science from other ways of acquiring knowledge or belief. It is said in favour of science and its method that they are peculiarly well suited to the end of knowledge generation. Tarot and tea-leaves are, in contrast with the scientific method, unreliable and cannot be used as a way of getting to know anything. Furthermore, the very superiority of the scientific method explains the great success of science. When western civilization hit upon the scientific method sometime in the sixteenth century (some say later) it discovered the key to understanding the natural world, which in turn allowed the manipulation of that world, for better or for worse, in hitherto unprecedented ways. So the existence of scientific method allows not only for the demarcation of science from non-science, but also for the explanation of the historical success of science. Popper, for instance, sees conjecture and refutation as providing not only a demarcation criterion – falsifiability – but also, under the description "critical rationalism", an explanation of the ability of science to progress.

This is what I shall call the "optimistic view". According to the optimistic view, scientific method has these features:

(a) it is distinctive of science – it characterizes science;
(b) it explains the success and progress of science;
(c) it is general in its application – it is common to all parts of science;
(d) it is applicable in a methodical fashion – it is not dependent on imagination, intuition, etc., and it can be applied by any reasonably skilled scientist; and
(e) its reliability can be known *a priori*.

I do not claim that many philosophers or others have subscribed to the optimistic view in every detail. What I want to show is that anything we might call a scientific method deviates considerably from the optimistic view. This will in turn allow us to see that there is nothing that can usefully be called *the* scientific method.

Rather, as we shall see, there are many methods used by science. Furthermore, the methods themselves are discoveries or applications of science. Hence they are not *a priori* methods. Some methods are more general and have a stronger *a priori* flavour. But it is difficult to call these "methods" because they are not sufficiently detailed to allow for a methodical application; they also tend to require the use of imagination and intuitive judgement. Although the view I will promote is not the optimistic one, it is not pessimistic either. Let us look first at proposals with an optimistic spirit.

The simplest proposal would be that of the naive inductivist:

> Go out and observe. When you see a well-established regularity, take that to be a law or causal connection.

How does this method rate as a candidate for the scientific method? This method has several virtues: it is general, it tells one what to do (look for regularities), and tells us what to infer (if there is a regularity, infer a law). However, the failings of the method far outweigh its virtues. Note, first, that its reliability, if it is reliable, cannot be known *a priori* – it is induction after all. Secondly, it is anyway not reliable. Many actual regularities are not laws (accidental regularities), and many perceived regularities are not actual ones (a summer's worth of observing green leaves). Thirdly, it is not unequivocally clear how one is supposed to use this method. We learned from Goodman's problem that with strange enough predicates anything could be made out to be a regularity. So the method should also tell us which predicates we should concentrate on. Fourthly, while it is quite general, this method is not perfectly universal – it does not help us infer the existence of unobservable entities. Fifthly, it does not fit the actual practice of scientists – they do not go out simply looking for regularities. Rather they may look for evidence for or against certain live hypotheses, or construct experiments to test them. And hypotheses are the products of conjecture and imagination.

This point suggests that the scientific method can have little or nothing to say about the invention of theories, the casting of hypotheses, and the dreaming up of potential explanations. That this must be so can be seen most directly in the case of discovering. Whatever the evidence there will be an infinite number of possible laws that will explain the evidence. There can be no method for generating all the corresponding hypotheses, and if there were it would be useless as we would have no idea which hypotheses to start testing. Writers on method therefore distinguish between the context of *discovery* and the context of *justification*, maintaining that scientific method concerns the latter not the former. The idea is that coming up with hypotheses is not subject to rules and methods, but that the testing, evaluation, and confirmation of theories is.

Popper and scientific method

This was Popper's view of method. Popper says that the process of coming up with a hypothesis – conjecturing – is an exercise of the imagination. But once a conjecture has been proposed, then it is clear what we must do. Now we must attempt to falsify it. We continue to do so until such time as a falsification comes our way, in which case we must draw the inference that the hypothesis is false. Apart from giving up on the process of discovery, this method is general and it is (supposedly) *a priori*. Falsification is also distinctive of science – it provides a demarcation criterion, and it explains the progress of science, which proceeds in a Darwinian fashion by eliminating falsified theories.

We have already seen that Popper's methodology cannot give us knowledge. Furthermore, it cannot work even on its own terms. Recall the cognitive dependence of observation on theory and the Duhem–Quine thesis. These suggest that it is impossible to make an observation that tests just the target theory. There will always be a host of auxiliary hypotheses, assumptions, supporting hypotheses, and so on, which may also be regarded as being under test too. A good example of this is the experimental disproof of parity conservation. Parity is essentially the left- and right-handedness of subatomic particles. For a long time it had been thought that parity is always conserved, meaning that physical processes should be symmetrical under spatial reflection. In 1956, T. D. Lee and C. N. Yang conjectured that parity is not always conserved in what are called "weak interactions". The symmetry requirement of parity conservation says that, when a nucleus decays, as many β particles should be emitted in the up direction as in the down direction. This was tested with the nucleus of a certain isotope of cobalt. A thin layer of cobalt-60 was polarized by a magnetic field so that the axes of the spinning nuclei were all in the same direction. The β emissions were measured in the relevant directions and an asymmetry was found, which falsified parity conservation. It is important that the cobalt nuclei are all nearly aligned. Otherwise the β asymmetry could be explained by supposing the nuclei to be pointing in different directions. How did the experimenters know they had succeeded in aligning the nuclei? They already knew that there is an asymmetry in the emission of γ particles by cobalt-60 and this could be measured. So they checked that this asymmetry was present while looking for the asymmetry in β emissions.

One of the auxiliary assumptions in this experimental test is the asymmetry of γ emission by cobalt-60. This is a highly theoretical general claim, as are the assumptions behind many other aspects of the experimental arrangement and the interpretation of the raw data. The Duhem–Quine thesis tells us that these are all under test together with the parity conservation principle. So, when we get a falsification, it is strictly the conjunction of all of these factors that has been falsified, not just a small part. A similar example might be the falsification of the claim that the solid state of a particular substance is always denser than the liquid state, which appears to be refuted by ice floating in water. But that observation may need to be tested, by checking that the ice and water are both uncontaminated. These tests will depend on general claims about the nature of water (e.g. its pH, boiling point, electrical conductivity). So the falsification is dependent on the assumption of the truth of these general assumptions. The only thing that we know is falsified is the conjunction of these assumptions and the hypothesis under test.

A Popperian might reply that at least something is falsified in such cases – the conjunction of hypotheses. But this move is only satisfactory if the conjunction is not too large. And we may argue that, for an anti-inductivist, this conjunction is very large – possibly extending to everything we believe. For among the propositions that will go into the conjunction will be all sorts of propositions which tell us that things are as we think they are and that nothing has interfered (e.g. "The experimenter did not commit a parallax error", "The experimenter had not taken hallucinogenic drugs", and so on).

Popper himself emphasizes a related point. Returning to the hypotheses and experiments just mentioned, the statements "The β emission showed asymmetry" and "The ice is floating in the water" are what Popper calls *basic* statements (and are what we would regard as observation statements, at least in the case of the floating ice). These are basic statements because, if they are true, they falsify the corresponding hypotheses. However, to assert such statements is to commit oneself to a range of general propositions: e.g. that γ emission asymmetry of cobalt-60 shows alignment of spin, that distortion of the magnetic field could not cause β emission asymmetry, or that at atmospheric pressure water freezes at 0°C and boils at 100°C, that the stuff under observation would boil at 100°C, etc. According to Popper, we can never know a general proposition to be true, and so we cannot know the basic statement to be true. This is known as the *problem of the empirical basis*. The problem of the empirical basis in turn raises a problem for falsificationism, for if we do not know any basic statements to be true, then we do not know that we have an instance that falsifies our hypothesis. The appeal of and motivation for falsificationism was the asymmetry between induction and falsification – while any number of basic or observation statements could not guarantee the truth of a general statement, just one basic statement can decisively show the general statement to be false. But this logical truth cannot be of any use to scientific method if it is also the case that we cannot ever know a basic statement to be true. We come again to the conclusion that falsificationism, if strictly anti-inductivist, is useless even in its own terms.

We saw in Chapter 5 that for Popper the decision to accept a certain weight of statistical evidence as falsifying a probabilistic hypothesis is a matter of convention. His response to the problem of the empirical basis is just the same. The decision to accept a basic statement as true is also conventional – a decision arrived at as a result of consensus among scientists. This establishes the basic statement as a sort of dogma, but a temporary and harmless one. This is because the basic statement is open to revision – were we to discover that this stuff does not boil at 100°C, we would revise the statement that we saw ice float on water. We saw that conventionalism was unsatisfactory in the probabilistic case, and the same is true here. Imagine we were to revise the basic statement because of some other discovery. That discovery would itself depend on certain basic statements ("The liquid boils at 95°C"), which themselves are accepted as a result of convention. Science, however much revised, will always rest upon convention. And there is nothing (non-inductive) to suggest that conventional decisions by scientists are true or likely to be true. And, in any case, if we are happy to allow (revisable) conventions to determine when a basic statement is true, why should not the (revisable) consensus of scientists also establish when a general statement is inductively confirmed by observations?

As Popper recognizes, even when we know what we have is a *prima facie* falsification, we do not always discard the theory. This is particularly true if there is no feasible alternative hypothesis. Scientists will continue to work with a falsified theory if there is no viable alternative on offer. Even if a pianist plays some wrong notes, if he's the only one in town, don't shoot him. Keeping Duhem and Quine in mind, we can always shift the blame onto one of the auxiliary hypotheses or background conditions (the equipment, the experimenter, bad luck, a bad day). However, this seems to make resisting falsification too easy. To keep hold of his method, Popper formulated rules that govern our response. An additional claim that saves the theory, for instance by saying that the background or test conditions are not as we supposed or by adjusting of an auxiliary hypothesis, is called an *ad hoc hypothesis*. Popper was not keen on *ad hoc* hypotheses, because they provide a let-out from falsification. He said that such *ad hoc* hypotheses should themselves be testable. This is certainly desirable, but it seems too strong as a requirement. At the simplest level it may be clear that one measurement out of several is a mistake – it is just too different from the others. Rather than consider the result as evidence (which would distort any statistical conclusions), it makes sense to discard the result. But it may be that the appropriate *ad hoc* hypotheses may be untestable – could we test whether the test tube used for an experiment carried out last week had not been washed properly? At the other end of the theoretical scale, the conjectured existence of the neutrino can be seen as an *ad hoc* hypothesis designed to square observed decay energies with the conservation laws. While there was eventually confirming evidence, at the time when Pauli first mooted the idea there was no conception of how this might be obtained.

Earlier we also saw that Popper held that science aimed at the increasing verisimilitude (nearness to the truth) of its theories. I argued that there is no non-inductive way of assessing the verisimilitude of a theory. Popper's conception of verisimilitude is nonetheless interesting, because it is influential in other conceptions of progress, in particular that of Lakatos. What then do we mean by saying that one theory is nearer to the truth than another? Verisimilitude serves to address the thought that two theories can both be false, but one can still be better, just as two arrows can miss the bull's-eye, but one can be a better shot by being closer to that bull's-eye. While we have an intuitive grasp of this notion, spelling it out in detail is difficult. One appealing thought is that the better theory will say more that is true than will the worse theory, as this example suggests. Consider the following:

> The density of mercury (at standard temperature and pressure) is 13.6 g cm^{-3}.
> Theory A says that the density of mercury is 13.4 g cm^{-3}.
> Theory B says that the density of mercury is 13.1 g cm^{-3}.

Clearly A is better than B, and although both are false both have true consequences, e.g. that the density is greater than 13.0 g cm^{-3}. Now consider the proposition *p*, that the density is greater than 13.3 g cm^{-3}. This true proposition is a consequence of theory A, but not of theory B; instead, theory B has the false consequence that the density is less than 13.3 g cm^{-3}. However, one cannot count the true and false consequences of the two theories, as both have infinitely many consequences, both true and false. Popper's idea is that we can nonetheless employ the notion of inclusion. Let us now look at all propositions of the form "The density of mercury is greater than x g cm^{-3}". When $x > 13.6$ g cm^{-3} the propositions are false and are consequences of both theories. When $x < 13.1$ g cm^{-3} the propositions are true, and again consequences of both theories. But when $13.1 < x < 13.4$ g cm^{-3} the propositions are true and are consequences of A but not of B. So, regarding claims of this form, all the true consequences of B are included among the true consequences of A, while A has some additional true consequences. Furthermore, all the false consequences of A (of this form) are also consequences of B, while A does not say anything false that is not also said by B.

Popper generalizes this idea, and defines comparative verisimilitude thus:

A has verisimilitude greater than or equal to B if and only if

> The true consequences of B are included among the true consequences of A; and the false consequences of A are included among the false consequences of B.

The picture of A having greater verisimilitude than B is shown in Figure 8.1 where A*t* is the set of true consequences of A, A*f* is the set of false consequences of A, and so on.

It turns out that this conception of verisimilitude cannot work. In the interesting case when both A and B are false and have different consequences, it cannot be both that the true con-

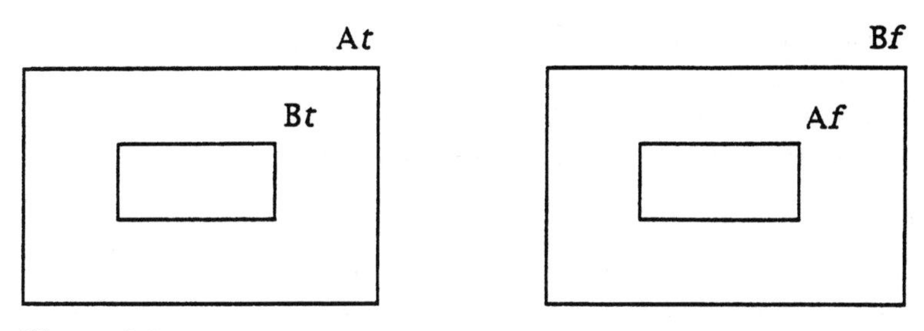

Figure 8.1

sequences of B are included among the true consequences of A and the false consequences of A are included among the false consequences of B. Let us see why this is. Thinking back to the theories about the density of mercury, consider

- *p* the density of mercury is greater than 13.3 g cm^{-3}; and
- *q* the density of mercury is less than 13.5 g cm^{-3}.

p is true and is a consequence of A, but is not a consequence of B. *q* is a false consequence of both theories. Now consider the conjunction of both these propositions, *p* & *q*. *p* & *q* is a consequence of A, as both *p* and *q* are consequences of A; but *p* & *q* is not a consequence of B, as *p* is not a consequence of B. Since *q* is false, *p* & *q* is a false consequence of A, but not of B. So the false consequences of A are not all included among the false consequences of B. But Popper's definition of verisimilitude required A to have no false consequences not already found among the false consequences of B.

The argument is easily generalized to show that, whenever two distinct theories both have some false consequences and some true consequences, they cannot be compared for verisimilitude.[54] Even the simple case we looked at above shows that Popper's conception cannot be applied to an obvious instance of one theory being closer to the truth than another. In which case it will not work for the sort of case it was conceived to deal with, for instance the greater nearness to the truth possessed by Einstein's theories as compared with Newton's theories. But, even if something like Popper's conception could work, it would still be of limited application, since it would not be able to compare theories which, although competing explanations of the same phenomenon, have quite different sorts of consequences. Most competing theories are like this. The competing theories of dinosaur extinction provide one instance. The meteorite theory will have both true and false consequences (e.g. concerning the existence and nature of impact craters) about which the volcano theory is completely neutral, and the same will be true the other way around. And so there is no reason to think that one theory will subsume the other (i.e. taking on all its true consequences). But that thought does not prevent us from thinking that one of these theories may well be much closer to the truth than the other.

We do have an intuitive idea of one theory being nearer to the truth than another. Why did Popper hit upon this ultimately unsuccessful conception of it, instead of some better one? One reason is an excessive concern for deductive logic. While deductive logic has held a powerful sway over much of philosophy, especially the philosophy of science, this is perhaps particularly true in Popper's case because of his insistence that it is the only form of legitimate reasoning in science. According to the logician's notion of a theory, a theory is a set of propositions that includes all the propositions which can be deduced from those propositions (what is called a deductively closed set). From this point of view it is natural to think of verisimilitude as being determined by the true subset of these propositions and the false subset of these propositions. In turn, theoretical betterness will be seen as shifting the balance between true set and the false set – adding to the true set and taking away from (or at least not adding to) the false set. This is the idea of theoretical inclusion we have seen in Popper's account of verisimilitude.

Briefly, I want to suggest that looking at verisimilitude from the point of deductive logic is highly misleading. Certainly theories aim at truth, but, with IBE in mind, we can see that they aim at truth by aiming to state the explanations of phenomena. In a rough sort of way we can draw a distinction between the general form of an explanation and its detail. So the meteorite and volcano explanations of dinosaur extinction differ quite markedly in their form, and both differ (to a greater degree) in form from the proposal that dinosaurs died out because evolutionarily superior creatures gradually supplanted them. However, two theorists who agreed that a meteor is the explanation but differ over its precise size would be differing in detail. While this distinction between form and detail is very imprecise, it serves at least to remind us that we regard a theory as getting near to the truth if it hits upon the right form of explanation, even if it gets the detail wrong. Say it really is the case that a meteorite caused the demise of the dinosaurs. Then the meteorite theory is nearer to the truth than the volcano theory. And that is true even if the size of the meteorite is significantly different from what the theorists estimated. Furthermore, we could imagine that, for some peculiar reason, the actual effects of the meteorite are in fact far more like the effects a volcano would have had than anyone supposed. Consequently, the volcano theory might fare better than the meteorite theory in terms of true consequences. But it would remain the case that the meteorite theory is better, as the form of the explanation is right, even if the detail misses the mark.

The point of the preceding comments is to suggest that a theory which gets the form right but the detail wrong might do rather badly by any logical/Popperian conception of verisimilitude. But we would still want to regard it as having got closer to the truth than a theory the form of which is mistaken. This is why we celebrate Copernicus' achievement in (re)introducing the heliocentric model of the solar system, even if, as historians have argued, his theory was less accurate than the best geocentric models. If I am right that the explanatory form of a theory is important in assessing the verisimilitude of a theory, then no purely logical account of verisimilitude will do. In particular, it will be wrong to think that, in order to be a better theory, a theory must include the true consequences of its rivals. The worse theory may still do better in some areas, by accident. This is important because, as we shall see, the spirit of Popper's approach has been adopted by his followers, notably Lakatos, and in consequence undermines their philosophies too.

Lakatos and the methodology of scientific research programmes

Popper's ideas were developed in a liberalized form by Imre Lakatos. Clearly one or two unexplained apparent falsifications –

anomalies as they are called – should not lead us to reject a theory. Nor should we be disturbed by the odd *ad hoc* hypothesis needed to prop it up. Nonetheless it is clear that something is wrong with a theory that accumulates anomalies and increasingly relies upon unverified *ad hoc* hypotheses. Lakatos was concerned to develop a scheme that would distinguish between theories which had a successful strategy for dealing with anomalies and those which did not. This was his *methodology of scientific research programmes.*

The theoretical part of a scientific research programme is composed of a *hard core* and of an *auxiliary belt*. The theoretical hard core is the definitive feature of the research programme. While the research programme as a whole may change over time, its hard core remains the same. What may change is the auxiliary belt. The auxiliary belt is what is needed in addition to the hard core to explain the observable phenomena. Let us say our research programme is that of Newtonian celestial mechanics (the favourite example for Lakatos' ideas). Our hard core will be Newton's laws of motion plus his law of universal gravitation. To apply this to the solar system we will need to use certain data or make various assumptions, for instance about the masses of the Sun and the planets and their location. We will also need to make approximations in order to facilitate calculations. Such assumptions will make up the auxiliary belt. These assumptions and approximations can be adjusted in order to make the hard core fit the observed motions of the planets. So, for instance, if we calculate the predicted motions assuming that the gravitational attraction the planets have for one another is negligible compared with the effect of the Sun, then we will observe a discrepancy between our predictions and the actual motions. Once attraction between planets has been taken into consideration this discrepancy can be eliminated. Another approximation leading to anomalies is treating planets as point masses. To account accurately for the motions of the Moon and Earth this approximation needs to be replaced by calculations taking into account the asymmetrical distributions of their masses. Doing so again eliminates the anomaly. A different case was the anomalous motion of Uranus. This problem was solved by hypothesizing the existence of an unobserved planet affecting the orbit of Uranus. This amounts to an *ad hoc* hypothesis. But it is in favour of the programme that the hypothesis was testable. The orbit of the unobserved planet was calculated, allowing it eventually to be found. This was the discovery of Neptune. (The white dwarf Sirius B, disturbing the motion of Sirius, was discovered in the same way, and for similar reasons a black hole in Cygnus X-1 has been hypothesized.)

Scientific research programmes have methodological components too – the *negative heuristic* and the *positive heuristic*. The negative heuristic is a stipulation that the hard core is not to be abandoned in the face of anomalies. The positive heuristic provides guidance on what to do in the face of an anomaly. In the cases considered, the injunction to improve on one's approximations is one way in which one could attempt to deal with deviations from the calculated predictions.

The question we asked was this: How can we tell whether a scientific research programme is doing well or faring badly? Lakatos gives three conditions. If these are met a programme is said to be *progressing,* otherwise it is *degenerating*. Over time the programme consists of a series of theories, each one superseding the earlier ones. As explained, all the theories will share the same hard core but will differ in their auxiliary belt. A progressive research programme has the following properties:

(a) later theories should have some content not possessed by earlier theories – they should say more;
(b) later theories should explain why earlier theories were successful; and
(c) later theories should have more corroboration than earlier theories.

Lakatos says that the negative heuristic, the decision to accept the hard core as true and to defend it against anomalies by adjusting the auxiliary belt, is conventional. A scientist who adopts the methodology of scientific research programmes must do this – it is part of the methodology. This seems to be conflating distinct things.

(a) *A new theory*. A scientist working with a new theory might not believe the theory to be true. It may simply have the status of a bold conjecture. The scientist does not abandon the theory in the face of anomalies, because the theory is in the early stages of development. If the anomalies continue to mount up and no attempt at adjusting the theory seems to make it much better at accommodating them, then the scientist might well give up the theory, having come to the conclusion that it is false. The scientist may never have believed the theory to be true, but may have hoped that it was. On the other hand, the scientist would not jettison the theory at the first obstacle. The theory may need some tinkering – perhaps the value of some constant is not quite right, even if the insight of the theory, say the general form of certain equations, is right. At this stage the relationship between the new theory and other theories, the experimental equipment, the observational background, and so forth, are not well understood, so the auxiliary belt needs to be developed as well. In these circumstances, not abandoning the theory does not amount to following a convention. It simply is what rational development of a conjecture amounts to. The scientist need not regard the theory as true to do this – she is just giving the conjecture its best shot. In the light of Inference to the Best Explanation, this is easily understood. Anomalies may be explained by the falsity of the theoretical core, but they may be also explained by all the other things that could go wrong (the equipment, the background assumptions, the values calculated for constants, etc.). In any case the scientist does not always refuse to countenance falsification of the theory – a young theory, which having been given its best shot is still beset with anomalies, may well be abandoned by its creator.

(b) *An advanced theory*. The situation is quite different for an advanced theory. A well-advanced theoretical hard core may well be regarded as true or containing a good measure of truth. But again this is not a decision governed by convention. Rather, such a theory is likely to be well supported by evidence and have considerable explanatory power. And that is why scientists believe it to be true – and it is also the reason why the theory is not abandoned in the face of anomalies. Past successes of the theory in dealing with anomalies might lead us to believe that a satisfactory dissolution of the anomalies will be found – that a neat adjustment to the auxiliary belt will be discovered which allows incorporation of the supposedly anomalous facts into the total theoretical structure (the research programme). And, in advance of such a resolution, it is likely that the mature theory will still retain much of its explanatory power. As long as it retains as much or more explanatory power than its rivals, it is rational to hang onto that theory. Again this is what one would expect in the light of Inference to the Best Explanation.

So its seems that the protection afforded to the hard core by the negative heuristic is neither absolute nor conventional. The negative heuristic is not an injunction against falsification. At most it sums up applications of Inference to the Best Explanation: (a) to develop a new and bold conjecture as far as possible before abandoning it in the face of anomalies – straightforward improvements in both the hard core and the auxiliary belt may remove (and so explain) the anomalies; and (b) if you have a powerful, well-tested, and well-corroborated theory that has dealt successfully with anomalies in the past, do not abandon it in the face of a new anomaly as new work may show how to deal with this anomaly too, and in any case, unless there is a very powerful rival theory on offer, this theory is likely to be the best and truest explanation of the phenomena you have got.

Lakatos describes the positive heuristic thus: it "... consists of a partially articulated set of suggestions or hints on how to change, develop the "refutable variants" of the research-programme, how to modify, sophisticate, the "refutable" protective belt".[55] He is not entirely clear as to how strong or contentful the positive heuristic is supposed to be. But some things he says suggest that the positive heuristic, whether articulated or not, directs the development of a scientific research programme in a pretty strong way. He says, "the positive heuristic sets out a programme which lists a chain of ever more complicated *models* simulating reality: the scientist's attention is riveted on building his models following instructions which are laid down in the positive part of his programme".[56]

Lakatos' critics have argued that positive heuristics do not exist in any but the most general forms. They might be injunctions like "seek unifying theories" or "prefer quantitative theories to qualitative ones". These do not give specific advice about how to adjust the theory in the face of an anomaly. To see why the positive heuristic cannot say more in detail, it is necessary to consider that there are at least two things we would like a theory or research programme to do:

(a) The theory should grow in breadth and in sophistication; and
(b) The theory should be able to cope with anomalies as they occur.

These goals are quite different. For it is in the nature of an anomaly that its occurrence cannot be predicted. On the other hand, early in the development of a theory there may be clear lines of research that need to be pursued – the application of the theory to phenomena hitherto not covered by the theory or making the theory more precise and less approximate. Examples of both are to be found in Lakatos' example of the progress of the research programme of Newtonian cosmology. Here the hard core comprises the laws of motion and of universal gravitation. In the first theory of the research programme the Sun and planets are treated as point masses, such that the mass of the Sun is so much greater than the masses of the planets that the gravitational effect of the planets on the Sun or on one another is strictly negligible. Newton would have known in advance that any fit between the predictions of such a theory and observation would only have been approximate at best, since he knew that the assumptions of this model are false. So the fact that the fit is only approximate does not count against the theoretical hardcore – as an anomaly might – but in its favour. Furthermore, aspects of the research programme are clear from the outset – the assumptions of this model, which are part of the auxiliary belt, need to be replaced by more accurate assumptions, i.e. that the gravitational effect of the planets on the Sun should be considered so that the Sun and planets rotate about their common centre of gravity, that the planets have a gravitational effect upon one another, and that the planets are oblate spheroids and not perfect and uniform spheres.

However, the failure of the fit of the orbit of Uranus was a genuine anomaly. The failures of fit hitherto considered could all be explained by regarding the existing model as just an approximation, where the nature of the approximation was known in advance. But this failure of fit could not be accounted for in that way, since it could not be known in advance whether there are unknown planets not taken into consideration. Of course, it could be considered in advance that there might be unknown planets that might generate anomalous observations. But any number of phenomena could be thought of in advance which might do so. It would scarcely be possible to list all of these, and in any case it would not be possible to assign a single cause to each conceivable anomalous observation. The task of constructing an algorithm which would set out precisely how to respond to any possible anomaly would be infinitely more difficult than working out what to do as and when an anomaly arises.

And so, insofar as the positive heuristic does give reasonably clear directions as to the development of a research programme, these do not cover those developments that arise as a result of anomalies. But it could cover those developments that are the improvements on acknowledged approximations in the model. Insofar as a positive heuristic could play a part in guiding the development of theory in response to an anomaly, it is only in the form of very general principles which might constrain development but not direct it, principles which need not be specific to that research programme.

What does this tell us about scientific method? If Lakatos were right we could adopt as our method the following: "Follow the positive heuristic of your research programme". But this depends on there being a detailed positive heuristic which gives reasonably precise instructions of how to develop the programme. With regard to improvements in the face of anomalies not due to acknowledged approximations, this simply is not possible; not every potential anomaly could be foreseen and a strategy for dealing with it planned.

Lakatos' methodology of scientific research programmes is also supposed to explain rational choice between competing programmes. When one programme is degenerating and its competitor progressing, it makes sense to switch allegiance to the competitor. Of course, in practice the choice of research programme by active scientists will depend on many other factors, especially professional ones. Nonetheless, all concerned should at least be able to see which programme is doing best, even if they continue to work on another one. Again this all depends on there being a reasonably detailed positive heuristic. (Even if it is only partially articulated, it may still be detailed.) For if a detailed positive heuristic is part of the programme, and the heuristic is what drives the development of the programme, then a failure to accommodate anomalies can be put down to a failure of the programme. However, if the positive heuristic has little detail, and the development of the programme is determined largely by external factors (the inventive genius of its proponents, the resources available to them, etc.), then the failure of the programme may be a result of these factors. Thus the lack of a detailed positive heuristic means that it need not always be the case that the degeneration of a research programme is evidence against it.

A more fruitful way of looking at matters is again in terms of Inference to the Best Explanation. The key factor in judging the merits of competing theories is their relative abilities to handle the currently available evidence. But other factors may be relevant and these may be diachronic (over time). So the recent accumulation of anomalies in a new area of research may be held to be the tip of an iceberg of failure. This will be reason to think poorly of a theory. But note that this is because future degeneration of a theory means the further accumulation of refuting evidence. Degeneration is not bad per se; what is bad is the evidence that degeneration brings with it.

My conclusion is that Lakatos' methodology of scientific research programmes, while it may provide some insight into the way that research programmes develop and compete, gives little assistance in the search for something that may reasonably be called "the scientific method". While Lakatos' descriptions are of a general nature and, in the Popperian spirit, have a strong *a priori* flavour, the prescriptive element lacks sufficient detail to be regarded as anything like a method. Furthermore, what Lakatos does say can best be understood as the application of Inference to the Best Explanation. Indeed, we shall see this in other cases of proposals for general methods of science. Later, I shall look specifically at the relation between inference to the best explanation and scientific method.

Clinical trials and the scientific method

Stuart Pocock, in an excellent book on the practice of clinical trials, refers to "the scientific method as applied to clinical trials".[57] He presents this diagrammatically as shown in Figure 8.2. Pocock then goes on to discuss in detail the five steps in the scheme and the procedures they involve. Two aspects are of particular interest. First is the statistical interpretation of the data and drawing inferences from it. We discussed this briefly in Chapter 6. Second is the emphasis on randomized, controlled trials. Controlled trials are those where the treatment is given to one group while another group remains untreated. The efficacy of the treatment will be measured by the difference in outcome between the two groups

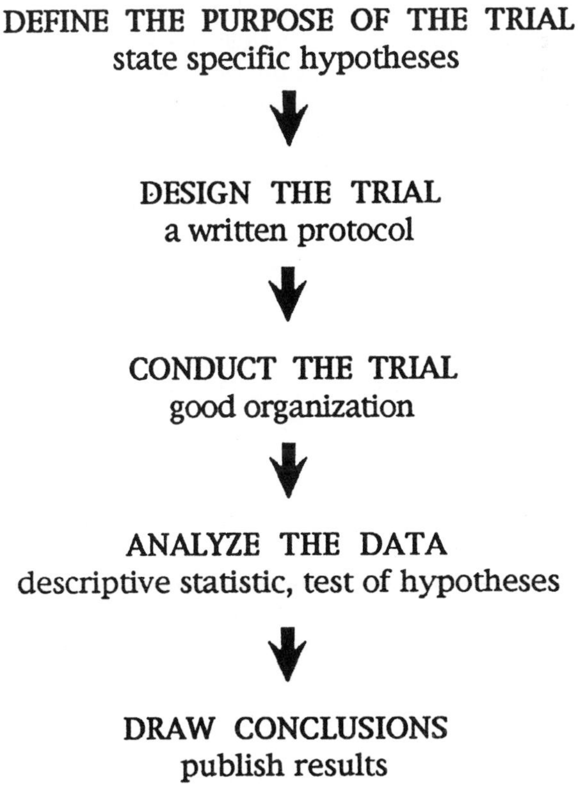

DEFINE THE PURPOSE OF THE TRIAL
state specific hypotheses

DESIGN THE TRIAL
a written protocol

CONDUCT THE TRIAL
good organization

ANALYZE THE DATA
descriptive statistic, test of hypotheses

DRAW CONCLUSIONS
publish results

Figure 8.2

(e.g. faster average recovery or lower mortality). Randomization refers to the fact that the allocation of patients to the control group or to the treatment group is random.

Pocock claims that "*randomized controlled trials are an essential tool for testing the efficacy of therapeutic innovations*".[58] It may well be true that the introduction of standardized procedures, including the techniques mentioned, has much improved the quality of clinical testing, particularly among those who are not full-time researchers. Furthermore, it is certainly very important in enabling the easy analysis of results by the doctors who read the many reports to be found in medical journals. However, taken literally this statement must be false. For, as Pocock himself says, randomized trials have only been in existence since the late 1940s. This would imply that therapeutic innovations had not been tested successfully before that date. But no one doubts that valuable research was carried out by many physicians prior to the 1940s. In the 1860s, for example, Joseph Lister experimented with carbolic acid to combat infection during surgery. His success in reducing mortality from 50 per cent to 15 per cent was taken, albeit after initial scepticism, as being a sure sign of the efficacy of antiseptic practices, even though his methods cannot be regarded as sufficiently rigorous by Pocock's standards.

Let us look at randomization. First note that randomization is not *sufficient* to ensure a satisfactory trial, even if everything else is as it should be. For a random assignment of patients may yield a very uneven distribution of patients (it is consistent with a random assignment that by chance all the oldest patients end up in one group and the youngest in the other – Pocock does not claim that randomization will ensure an unbiased assignment and acknowledges the point). For similar reasons randomization is not *essential*, for one can imagine perfectly reasonable non-random procedures giving the same assignment. There are two sorts of bias that randomization is supposed to eliminate. First is selection bias – for instance the optimistic researcher may consciously or unconsciously choose those patients for treatment who seem healthier. Second is bias resulting from inherent differences in patients, whether visible or invisible, such as age and psychological state, which might affect the outcome. Randomization is supposed to eliminate these effects by distributing them evenly over the control and treatment groups. As regards selection bias, this could, in principle, be eliminated by unrandomized double-blind testing, e.g. where drawing up of the groups is done in a non-random way and the decision about which group should receive the treatment is made afterwards or by some independent person. For the purpose of evening out differences between the two groups, informed deliberate distribution will do just as well as randomization. For instance, if age is thought to be a relevant factor, then it must be ensured that the two groups have similar age profiles. This will do at least as well as randomization. Of course there may be factors of which we are ignorant. But these will, in effect, be randomly distributed by any method of assignment. A simple example from the origins of randomization will do. Fisher developed the process of randomization in connection with agricultural trials. Randomization would ensure that there is no bias such as allocating plots with better soil and conditions to the new variety under test rather than to the old control variety. Farmers and researchers know that some parts of a field have better drainage, more light, less wind, and so on. An unrandomized way of overcoming such potential distortions would be to divide the field in question into a chequerboard pattern, alternating the variety between plots. This way the condition of one plot will not be too different from that of its neighbours.

Reference is often made to *double-blind* trials. In such trials neither the patient nor the clinician knows whether the patient is receiving the treatment. This is often done in conjunction with the use of a placebo, which is a harmless substitute equivalent in appearance to the treatment. The point of double-blind trials is to eliminate two effects: (a) the very fact that a patient is receiving what he or she takes to be a treatment can have beneficial psychological and psychosomatic effects; and (b) if the clinicians know which patients are receiving the treatments there may be a tendency to treat them differently from those without the treatment and thereby bias the results. Clearly, however, double-blind trials are not always appropriate or possible. For instance, new surgical procedures cannot be done without the knowledge of the surgeon. Nor would it be right to open up control patients under anaesthetic, which would be the equivalent of giving them a placebo in a drug trial.

In practice it probably does turn out that randomized controlled trials, double-blind where possible, are the best way of testing new treatments for efficacy. Pocock and others provide plenty of corroborating evidence. Yet there seems to me to be nothing *a priori* about this. Experience has taught us that this way of doing things is effective. And we may learn better ways of doing things yet.

A spectrum of methods and principles

Lakatos' methodology of research programmes provided a very general *a priori* description of how one should pursue a research programme and evaluate it against its rivals. At the same time it could not be said to provide anything like a method. We saw that there simply is not enough detail for that. Handbooks for clinical trials provide much more detail and can be followed in a methodical fashion. Even so, much is left to the judgement of the clinician (for instance whether a double-blind trial is feasible or desirable). Such trials do not in any case have universal application – they are limited almost exclusively to the biological and medical sciences. Furthermore, the principles of clinical testing have a large *a posteriori* component. Another method, the role of which is to prove that a certain organism causes a certain disease, consists of Koch's postulates. This method involves: (a) culturing the organism from a diseased animal, (b) growing a pure culture, (c) causing the disease in a healthy animal from the pure culture, and (d) isolating the same organism from the animal thus infected. If these things can be done, then we have found the cause of the disease. Koch's postulates provide a methodical procedure for providing knowledge of a specific sort. Clearly they are not *a priori* but are themselves the product of microbiological knowledge.

Getting more methodical still are the many tests and other methods used in science. For instance the Wasserman reaction, which detects syphilis, or the eponymous litmus test. These can be applied almost mechanically to provide scientific knowledge of an albeit limited sort. These are certainly scientific methods, but scarcely *the* scientific method. Their roles are specific. Furthermore, they are patently *a posteriori*.

What conclusions may we draw so far about the scientific method? I can find little concurrence as to what it might actually be – the reason being, I conclude, that there is no such thing. If the scientific method is a *method for producing scientific knowledge* then there is nothing that is both a method and has sufficient generality to be called *the* scientific method. Instead I suggest that there is a spectrum of methods, rules of thumb, general methodological principles, and heuristic precepts all of which play some role in generating scientific knowledge. At the one end of the spectrum we have, for example, Lakatos' remarks on the methodology of scientific research programmes. These have a strong *a priori* flavour to them. While Lakatos does detail the sort of considerations which are relevant to assessing the relative merits of research programmes, he scarcely gives rules for generating such an assessment. At the other end of the spectrum are genuine methods, but the application of which is limited in scope and the efficacy of which can be known only *a posteriori*, being the products of specific scientific discoveries.

One reason why some aspects of scientific activity cannot be regarded as subject to scientific method is that they lie in the realm of genius, imagination, and invention. Famously, Kekulé's discovery of the benzene ring was said to originate in a dream he had of a snake biting its own tail. Method cannot account for that, nor could it substitute some alternative, methodical route to the production of new ideas. Serendipity, chance, dreaming, even tarot and tea leaves cannot all be eliminated from the scientific enterprise. This is not to say that discovery need always be entirely ruleless happenstance. Mill's canons provide useful guidance in the search for causal connections. For instance, the method of difference says that if circumstances A result in effect E but circumstances B do not, then if A and B differ only in that A includes feature C but B does not, then C is the cause of E. Clinical trials seek to approximate the method of difference. However, the method of difference is limited to cases where the subject matter can be controlled to eliminate all differences but one – hence the difficulty in forming and testing causal hypotheses in sociology. And, in any case, the application is typically approximate. In a clinical trial not every difference between control and treatment groups can be eliminated. Researchers have to use their judgement as to whether certain differences are likely to be relevant or not (e.g. the star signs of the patient's paternal grandmothers). The conclusion is that the generating of hypotheses cannot be done methodically.

The context of discovery and the context of justification

It is because of this that a distinction is made between the context of discovery and the context of justification. The former refers to the process of devising hypotheses, while the latter refers to their evaluation. Claims for the existence of the scientific method would therefore seem best limited to the context of justification. However, it is no more clear that these (the testing, evaluation, and confirmation of theories) can be specified or carried out in a purely methodical manner. The devising and carrying out of novel experiments requires considerable inventive imagination.

An illustration of this is the experimental resolution of the debate over the discreteness of electrical charge. Robert Millikan thought that the electron carried the basic unit of charge, while Felix Ehrenhaft considered that charge could come in continuous quantities. How could one test such hypotheses? Millikan's famous oil drop experiment involved the spraying of oil drops in such a way that they would become charged. The terminal velocity of such a drop falling through an electric field could be measured. This gave a measure of the charges on the drops. The fact that the measured quantities were multiples of a single unit confirmed Millikan's view against Ehrenhaft's.

There is nothing about the hypotheses under debate which suggested this particular experimental test. Even carrying out the experiment and obtaining data from it are not methodical matters. Millikan has famously been accused of fudging the results. It is clear that he discarded "bad" results. It may be said, perhaps, that if this is true then Millikan was not properly following the scientific method; following the scientific method would require faithfully recording every result. This is too simplistic. Experiments of this sort are not like making a litmus test. They involve new equipment, new techniques, and new calculations. Any of these, including the calculations, can go wrong. An experimenter would not expect to get useful data straight away. The apparatus will need fine-tuning and other adjustments. In getting the set-up to work and in making the observations and measurements the experimenter will get to know the experiment, for instance how sensitive it is to laboratory conditions. The experimenter will acquire certain skills in getting the experiment to work and in detecting whether it is indeed working or whether conditions are disadvantageous. So the fact that Millikan threw out some results is not enough to convict him of unscientific conduct. It may be just that he was a very good experimentalist (and he was) and therefore was able to tell usable results from unusable ones. Of course this can shade unconsciously into bias. But even sometimes bias itself may reflect a good nose for the truth. After all, when a student carries out the oil drop experiment, the result obtained is neither a confirmation nor a disconfirmation of our beliefs about electronic charge. Rather it is a test of the student's competence as an experimenter. What is true of today's student may also in part be true of Millikan himself. Given what Millikan had reason to believe about electrons and given the results he had already got, Millikan might have been justified in throwing out anomalous data on the grounds that the best explanation of it was that the experiment was playing up.

If the design, conduct, and interpretation of experiments is also a matter of imagination, skill, and intuitive judgement, then perhaps what we mean by the context of justification should be limited solely to the assessment of evidence. Yet even this is difficult to demarcate clearly. Say the evidence presented arises from a novel experiment. If those assessing the evidence have not themselves carried out the experiment then they may be unsure of the true significance of the results. Do the results really have the theoretical implications that they suggest? Or might they be peculiarities of the apparatus or of other features of the experimental context (including the experimenters themselves)? This is one reason why scientists regard the replicability of experiments as important. To assess experimental data fully a scientist will want, if possible, to repeat the experiment that generated them.

A notorious example of the role of replication was the cold fusion affair. In 1989, Martin Fleischmann and Stanley Pons, chemists working at the University of Utah, announced that they had demonstrated the low-temperature fusion of deuterium (an isotope of hydrogen) to form helium. They claimed that the heat and neutron radiation generated by their apparatus could only be satisfactorily explained by such a fusion process. Because cold fusion was potentially a remarkable source of energy, scientists from around the world immediately took a great deal of interest in the results. Many laboratories sought to repeat Fleischmann and Pons' experiments. Early on several reported confirmation of the observations. Texas A&M University reported excess heat in their experiments, Georgia Tech detected neutrons, and other similar reports were heard of from other groups. But, eventually, as these and other researchers began to understand the apparatus better, they retracted their claims. Georgia Tech's neutron detector was sensitive to heat, and the wiring of a temperature-measuring device explained the results obtained at Texas A&M. Eventually, most scientists concluded that, whatever results may have been found, they were not to be explained by cold fusion, at least not on a significant scale.

What repeating the experiment will allow the scientist to do is to see whether the hypothesis under test really is likely to be the best explanation of the results obtained. In the cold fusion case two groups concluded that their positive results had nothing to do with cold fusion – they had alternative explanations. This shows that in the actual practice of science the assessment of experimental results, and so their use in justifying a scientific claim, can require similar non-methodical skills and judgement that are required in the devising of the experiments themselves.

This case leads to a more general point about the assessment of evidence. According to Inference to the Best Explanation, the assessment of a hypothesis against the evidence is not simply a matter of seeing whether the evidence is plausibly explained by the hypothesis, but also of considering whether there is reason to think that it is the best explanation of the evidence. To do that it will be necessary to conceive of what the alternative explanations might be. But conceiving of possible explanations is part of what we called the "context of discovery", which we said was not amenable to method. It is not always possible to distinguish the context of discovery from that of justification. Although the distinction may be a useful one, the fact is that for most of science the two are bound up together; the one requires the other.

Inference to the Best Explanation and scientific method

Throughout this book I have suggested that if there is any overarching principle upon which scientific reasoning works it is Inference to the Best Explanation. Does not Inference to the Best Explanation constitute the scientific method? I do think that the central role of Inference to the Best Explanation does explain many of the proposals for the methods of science we have discussed. Insofar as Lakatos is right that we look unfavourably upon a degenerating research programme in comparison to one that is progressing, it is because the signs of degeneration (increasing use of *ad hoc* hypotheses, accumulating anomalies, increasing complexity, and so forth) are all signs of a poor explanation. Recall the method of clinical testing. Here the design of clinical tests seeks to ensure that the only plausible explanation (and so the best explanation) of a difference between a test and a control group is the efficacy of the treatment. The relation with Inference to the Best Explanation also explains other proposals for methods in science. For instance Bill Newton-Smith lists eight "good-making features of theories" which guide us in theory choice:[59]

(a) *Observational nesting.* A theory ought to have at least the same observational consequences as its predecessors.
(b) *Fertility.* A theory should allow for further research and theoretical development.
(c) *Track-record.* A theory should have had successes in the past.
(d) *Inter-theory support.* A theory should integrate with other theories.
(e) *Smoothness.* The theory's failures should be systematic, allowing for a smooth adjustment to the theory to rectify it.
(f) *Internal consistency.* The theory should not be self-contradictory.
(g) *Compatibility with well-grounded metaphysical beliefs.* The theory should be consistent with certain general or metaphysical assumptions about the world.
(h) *Simplicity.* Simple theories are preferable to complex ones (Newton-Smith himself thinks this is merely a pragmatic preference).

To these we might add other desiderata, such as:

(i) *Quantitative predictions.* A theory that provides quantitative testable predictions is preferable to one that is merely qualitative.
(j) *Novel predictions.* A theory that makes unexpected but corroborated predictions is preferable to one that does not.

This list is not exhaustive. Nor are these guidelines exceptionless. A theory may not maintain all the observational successes of its predecessor if it has significant successes elsewhere. So the heliocentric view of the solar system superseded the geocentric one, despite the fact that the latter, but not the former, seems to explain successfully why we only ever see one face of the Moon. Similarly, the theories of relativity ousted Newtonian mechanics despite a loss of simplicity. The criteria given are features which, other things being equal, we prefer a theory to have. As we have seen, they may on occasion compete and judgements will have to be made in particular cases as to which features are more important. Often one may think that there are insufficient grounds for a decision and that further theoretical development and testing are required.

The desirable qualities of theories given above can all be subsumed under Inference to the Best Explanation. Let me show how:

(a) *Observational nesting.* If one theory explains all the observations another explains, and more besides, then, other things being equal, it is a better explanation.
(b) *Fertility.* Why should we prefer more fertile theories? One reason is simply that a fertile theory is more interesting to work with. But there is more to it than this. If a theory is fertile it will suggest novel potential observations or links with other theories. These provide opportunities for making the theory a better explanation. An infertile theory, by contrast, offers little scope for improvement.
(c) *Track-record.* Clearly this is the key feature – the theory has to be a successful explainer of the facts.
(d) *Inter-theory support.* In Chapters 1, 2 and 4, we saw how mutual integration is part of our concepts of law and

explanation, and so mutual integration among theories is a sign of their truth.

(e) *Smoothness*. An explanation may not get everything right, but the best explanation of the fact that the exceptions are systematic is that our theory is right in some important respects. The ideal gas laws are successful for certain ranges of temperatures, pressures, and volumes, but break down for low temperatures and high pressures. This fact is best explained by taking the gas laws to be near to the truth – an approximation that gets worse as certain factors which have been left out of the theory become significant at low temperatures and high pressures (attractive forces between molecules).

(f) *Internal consistency*. If a theory is self-contradictory it cannot be true and *a fortiori* cannot explain anything properly. Clearly the removal of the inconsistency is important. Having said that, we have just seen that false theories can provide approximate explanations, and the same may be true of strictly inconsistent theories (inconsistencies in the calculus did not prevent its successful application to many scientific problems in the seventeenth and eighteenth centuries).

(g) *Compatibility with well-grounded metaphysical beliefs*. It is not always clear what the nature of these beliefs is. Nonetheless, if we think that certain general propositions are well established, then they will put constraints on what counts as an acceptable explanation. For instance, many are sceptical of the claims of homoeopathy since it is not clear how a treatment which is so dilute as not to contain any active ingredient could bring about a cure.

(h) *Simplicity*. Again, as we have seen, simplicity along with systematic integration is a criterion of lawhood and so of nomic explanation.

(i) *Quantitative predictions*. Precise quantitative predictions have two features. They increase the scope for falsification, and so a successful quantitative prediction rules out a greater range of competing hypotheses. Furthermore, non-quantitative, qualitative success may be more open to alternative explanation in terms of fudging.

(j) *Novel predictions*. Any successful prediction favours the theory which made it. This is because the more a theory explains the

> better an explanation it is. Novel and unusual predictions are of the greatest value since they provide instances explained by the theory in question but not by its competitors. If the predictions come before their confirmation, then this also rules out fudging as the explanation of success.[60]

There is reason to believe that if there are things which can be called scientific methods, then the more general ones can be seen as aspects of Inference to the Best Explanation. At the same time, inference to the best explanation allows us to see why the optimistic assumptions about scientific method cannot be fulfilled. As I remarked in the previous section, to infer the best explanation we need to have some idea what the competing explanations might be. Someone who does not have the experience or imagination to conceive of alternative hypotheses, but instead latches on to the first they hear of, is unlikely to succeed in inferring true explanations. By contrast, the scientist who sees what plausible competing explanations there may be of the data, is more likely to infer correctly when they infer from the best of them. But thinking up plausible hypotheses was something we ascribed to the context of discovery, which may not be amenable to method. There cannot be a systematic or *a priori* way of generating possible explanations. And so inference procedures that are cases of inference to the best explanation cannot be examples of the use of "the scientific method". As such procedures generate most of the interesting scientific knowledge there is, that knowledge is not the product of method.

Science without the scientific method

To a large extent my view concurs with the arguments given by Paul Feyerabend in his book *Against method*. There he seeks to show that no general methods, principles of inference, or rules of investigation are exceptionless. For various plausible proposals for such principles, he shows that there are instances where science has been advanced by going against them. From this Feyerabend concludes that no method should be discarded by science. Infamously, he concludes "... there is only *one* principle that can be defended under *all* circumstances and in *all* stages of development. It is the principle: *anything goes*".[61] He lays down the challenge that the methods of voodoo and witchcraft should therefore be accorded respect too. It has generally been felt by both supporters and opponents of Feyerabend that, if his arguments about method are correct, then the consequences for the traditional picture of scientific knowledge are negative. We would have to abandon the view of scientific knowledge as growing cumulatively and as representing a special cognitive achievement of the modern world. I suspect that it is felt on both sides that such claims for science depend upon the existence of the scientific method. Therefore if science needs to be defended, the notion of scientific method needs defence too. Correspondingly, if the optimistic view of scientific method is discarded, then we must let in voodoo, witchcraft, and no doubt reading tea leaves and crystal ball gazing too.

Here I disagree. The defender of the traditional view of science can grant that there is no scientific method. Let us recall the lessons of reliabilism. These were that, for knowledge, it is sufficient that our beliefs are true and produced by reliable methods. Furthermore, we saw that such methods themselves are open to investigation, discovery, adaptation, and improvement. The development of such methods itself is the product of science. There is no need to maintain that there is a single, exceptionless, and *a priori* method that is responsible for the growth of scientific knowledge. Taken in this sense we can endorse the conclusion that the only principle that comes close to this status is *anything goes*. Independent of knowledge altogether we cannot reasonably either establish or reject methods.

Does this require that voodoo and reading of tea leaves be taken seriously? Feyerabend was not suggesting that voodoo practitioners be set up in research centres with massive budgets like the CERN particle physics laboratory. After all, the purpose of voodoo is not primarily epistemic but religious. (Voodoo mixes features of Catholicism and of West African religions.) Nonetheless, adherents of voodoo do claim certain sorts of knowledge. We do tend to reject such claims, since they neither originate from standard science nor are easily accommodated by it. Worse, we tend to reject voodoo without knowing much about it. If one takes the optimistic view of method, one may well be sceptical about the possibility of sophisticated knowledge gained without it. But if one rejects the optimistic view and replaces it with the reliabilist approach, we should be more open to the possibility that other cultures may have developed or stumbled across methods that happen to be reliable. Furthermore, it is perfectly compatible with a method being reliable that those using it misunderstand it. For instance, a witch doctor may have a procedure for making prognoses on patients with a certain illness. He attributes the success of the method to occult powers, spirits of ancestors, and the like. Yet the procedure may be reliable for physiological reasons unknown to him. Such ignorance does not prevent him from using it to make successful predictions about recovery or death, predictions which can amount to knowledge. In this way reliabilist epistemology is more liberal than internalist epistemology or the optimistic view of scientific method.

This should make us want to modify the cumulative/progressive, traditional picture of scientific knowledge. First of all, science cannot have an exclusive claim to (sophisticated) knowledge. Secondly, the accumulation of knowledge is imperfect. Some pieces of knowledge get forgotten if they are not accommodated. It is probably the case that many efficacious herbal remedies have been ignored or lost altogether because they lacked the prestige of scientific medicine.

Thus qualified, the traditional view remains largely intact. On the whole, scientific knowledge is largely cumulative. Thanks to science, we know a lot today that we did not know a hundred years ago, and much of this knowledge has been built upon things discovered earlier. This is because the history of science is not just the acquisition of knowledge but also the use of that knowledge to forge new methods for gaining knowledge. It is because such methods are the product of science that Feyerabend is right that there is no one general method which will yield scientific knowledge, whatever our stage of scientific development.

The traditional view asserts that our scientific beliefs are largely true and reliably produced. Do we have any reason to believe this? Referring back to the beginning of the chapter, such an issue is an *a posteriori* one. A philosopher cannot give an *a priori* answer to it. Perhaps one reason for the opinion that a justification of the scientific method must be *a priori* is the thought that the issue of the progress of science should be a philosophical one.

However, it is not. But the philosopher can say what sort of consideration would decide the matter. What would show that our scientific beliefs are, by and large, true and reliably produced? The interesting bit is showing that the methods are reliable. For, if that is the case, then by the definition of reliability the beliefs produced by the reliable methods would indeed be largely true. In any case, if we need reasons to think our scientific beliefs are true, then the evidence and justification offered by scientists are the best we can hope to get. As I explained in Chapter 4, there are no special philosophical arguments outside science that establish the truth of scientific theories. The key question is whether the methods, inference patterns, and so on, which lead scientists from this evidence to their conclusions are reliable.

We have seen that the methods by which we generate our scientific beliefs are not *a priori*. At least they are not entirely so, in the sense that, although *a priori* methods (e.g. mathematics) are used typically these *a priori* methods are used in conjunction with *a posteriori* ones. That being so, it will again be a non-philosophical question whether the methods we use in science are reliable. Indeed, it will typically be a scientific question.

For instance, is the litmus test a reliable way of telling whether a solution is acidic or alkaline? Is carbon dating a reliable way of discovering the age of wooden artefacts? The answers should come from a chemist or physicist, who would tell us that they are. How do they know? They might tell us about the different buffers that make up the litmus solution and explain how they work, and, if pressed may tell us what the evidence behind such knowledge is. They may also be able to tell us about the half-life of carbon-14, and so on. At the level of tests that are able to give us quite specific knowledge, we are often able to explain their reliability. Typically this is because they have been devised with the relevant knowledge in mind, but in medical cases diagnostic tests may be based on well-established correlations which only became understood at a later date.

Were the optimistic view of method right, it would have this advantage, that we would be in a position to tell that the scientific method is reliable, and hence that science does consist of knowledge which accumulates and progresses over time. To accept the continuum view is to give up the possibility of a straightforward assessment of the reliability of the formation of scientific belief. Many among these methods may be unreliable. Certainly some possible methods are unreliable. Obvious cases include the examination of the entrails of sacrificial animals and phrenology. A scientific investigation might tell us whether the methods of witch doctors can be relied upon or not. Other methods may seem (to us) to be more plausible, yet are not reliable. Alan Chalmers cites Galileo's method of determining the ratio of the diameter of a star to its distance from the Earth. He did this using a taught vertical thread.[62] Galileo would then move himself until the thread exactly obscured the star. The ratio of the thickness of the thread to the observer's distance from it should be equal to the ratio of the star's diameter to its distance from the Earth. However, as Chalmers points out, this cannot work because of the distortion caused to the star's image by refraction in the Earth's atmosphere. The use of many methods of discovery requires skill and judgement on the part of the experimenter – Millikan's method for determining the charge of the electron was a case of this. On occasion this skill can be lacking or the experimenter's judgement clouded. The case of Blondlot's N-rays exhibits this. Blondlot thought he had discovered a new sort of ray, along the lines of cathode rays and X-rays. But it turned out he was deceiving himself when he continued to "observe" the rays when a crucial prism had been removed from his apparatus. In other cases equipment is designed with faulty assumptions on the part of its designers. The British team who discovered the ozone hole above the Antarctic for a long time did not believe their results. This is because they conflicted with the results of the American team, which had much more sophisticated equipment. It turned out that this equipment had not detected the ozone hole because it had been programmed to ignore large deviations from the norm in measured ozone, on the assumption that large deviations could only be the result of errors.

Yet I think that thoroughly unreliable methods are the exception rather than the rule. Such unreliable methods tend to get discarded, although a method that is unreliable at the margins may be retained. Recalling the discussion of knowledge and reliability, it is sufficient for knowledge that our beliefs are supported by reliable methods. For such methods to produce knowledge it is necessary that they be reliable, but not that we know them to be reliable. Yet these methods may themselves be investigated and their shortcomings and successes analyzed and the efficacy of the method improved – and the same goes for the knowledge, skill, and inventiveness required to apply the methods. Thus as science grows so will the number and range of methods available, and their limitations will be better understood. This in turn will assist in the greater production of scientific knowledge, which feeds back into our understanding of method. It is this feedback, whereby the increase in science means an increase in the methods for producing science, which explains the exponential growth in science.

Method and the development of science

Does the lack of a single method of science discredit science and the claims made on its behalf regarding rationality, knowledge, and progress? If there is no such thing as the scientific method, can we make a distinction between a scientific investigation of a subject and a non-scientific investigation? I think it is still possible to do so. Even if there is no one thing that is the scientific method, there are still scientific methods. As I have maintained in the previous section, there are methods that science itself tells us are reliable. Science, experience and reason tell us that the method of clinical testing, even if not perfect, is a highly reliable method of assessing the efficacy of treatments. So using a clinical trial will be to make a scientific investigation of a scientific claim. Reading tea leaves, however, is unscientific. There is no theoretical reason for thinking it is likely to be effective. Nor does experience tell us that it is – there is no inductive evidence for thinking reading tea leaves is reliable.

There is another aspect of a scientific investigation which will differentiate it from the unscientific. This lies in the range and kind of potential explanations which are considered. In the Introduction, I described how Judge Overton declared that scientific explanations involved natural laws and the forming hypotheses that can be tested against observation. This contrasts with the ready recourse of the creationists to supernatural explanations. Hempel cites an example of a non-scientific explanation from the astronomer Francesco Sizi:

> There are seven windows in the head, two nostrils, two ears, two eyes and a mouth; so in the heavens there are two favourable stars, two unpropitious, two luminaries, and Mercury alone undecided and indifferent. From which and many other similar phenomena of nature, such as the seven metals, etc. which it were tedious to enumerate, we gather that the number of planets is necessarily seven ... Moreover, the satellites are invisible to the naked eye and therefore can have no influence on the earth and therefore would be useless and therefore do not exist.[63]

Hempel says that this is an entirely spurious explanation. I am not so sure. If one were convinced that the mind of God the creator delighted in such symmetries, then this may well be a potential explanation. But it is not a scientific one, since it is not natural law that is being appealed to. Arthur Koestler argues that even Copernicus was wedded to non-scientific forms of explanation. For instance, in order to explain why the rotation of the Earth does not destroy it (by radial, i.e. centrifugal and centripetal, forces), Copernicus appeals to an Aristotelian distinction between "natural" and "violent" motions:

> But if one holds that the earth moves, he will also say that this motion is natural, not violent. Things which happen according to nature produce the opposite effects to those due to force. Things subjected to violence or force will disintegrate and cannot subsist for long. But whatever happens by nature is done appropriately and preserves things in their best conditions. Idle, therefore, is Ptolemy's fear that the earth and everything on it would be disintegrated by rotation which is an act of nature, entirely different from an artificial act anything contrived by human ingenuity ...[64]

Copernicus is arguing that there is a difference between the rotation of the Earth and the rotation of a bucket around a person's head – the difference is that one is natural and the other not. Clearly it is not a good explanation (after all, natural processes, such as volcanoes and hurricanes are destructive). But, more importantly, it is the sort of explanation which is of dubious scientific value. It does not appeal to universal natural law; on the contrary, it appeals to an old Aristotelian distinction (natural vs unnatural), which plays no part in scientific thinking (according to which all phenomena of a certain kind are subject to natural law, irrespective of their origins).

Galileo, by contrast, when faced with this and similar objections, responds by bringing diverse phenomena under the same explanation. The fact that a falling object does so in a straight line downwards seems to contradict the claim that the Earth is moving. If the latter were the case one would expect the line of fall to be an arc, as the Earth moves under the falling object. The object should get left behind and fall some way to the side. Galileo first points out that we can only tell whether the line is straight on the assumption that the Earth is stationary, which is the point at issue. He then demonstrates that we should expect that terrestrial bodies share the same motion as the Earth, pointing to an acknowledged phenomenon demonstrating the same principle. This is the famous example of the cabin in a moving ship, in which living things (flies buzzing around and fish in a tank) and non-living things (water dripping from one bottle to another) behave just as they would if the ship were stationary.[65]

It is clear that some significant changes in thinking allowed for the great blossoming of scientific thought at the end of the Renaissance. Sometimes this is accounted for in terms of the discovery of the scientific method. I think this is misleading. First, I have argued that there is no such thing as "the scientific method". Secondly, if the discovery of some perfectly general scientific method were the appropriate explanation of the scientific revolution, one might expect the simultaneous development of all sciences. Yet significant advances in chemistry and the life sciences had to wait until the late eighteenth and nineteenth centuries. I think there must be another explanation, or, more likely, set of explanations. One hypothesis is that the kinds of explanation that thinkers employed were changing. I suggest that there was emerging a clearer conception of natural law, and that explanations were increasingly of the form discussed here in Chapter 2, in contrast to the sort of explanations employed by Sizi and Copernicus, where instead of laws of nature we have the psychology of God or vague principles of limited or unclear application. In particular, the explanations began to take on a precise mathematical form, allowing for quantitative prediction and confirmation. Along with this went a waning trust in ancient principles, results, and observations. According to Koestler, Copernicus is an interesting case here, for he depended largely on the observations of his esteemed predecessors, such as Ptolemy and Hipparchus, denouncing those who questioned the latter thus "... it is fitting for us to follow the methods of the ancients strictly and to hold fast to their observations like a Testament".[66] Galileo said that he found no reason to take Archimedes (with whom he agreed) to be a greater authority than Aristotle (with whom he disagreed), except where the propositions of the former were supported by experimental evidence.

Roughly speaking, inferences using explanation can go in opposite directions. For instance, one can use a knowledge of laws and general facts to explain or predict observable events. Or one can use observational knowledge to infer the existence of laws and causes that best explain them. I conjecture that one of the changes wrought by the Renaissance regards the balance between these two directions of inference. To the medievals what was most certain was the nature of God as laid down by the Church and scripture. From the nature of God, philosophers sought to infer principles governing the world He created. These in turn could be used to explain observed phenomena. Since theology gives us little reason to think of God's nature in precise quantitative terms, the inferred principles would themselves be imprecise and qualitative. It is easier to fudge an explanation in such terms than it is where the phenomenon and the explanation are described quantitatively. During the Renaissance greater emphasis was laid on the reverse pattern of inference – from observations, to laws which would explain them. Not that God was excluded, for the extension of this pattern of inference was to draw conclusions about the nature of God, which gave rise to natural theology – understanding God through what we learn by observation of His creation.

While it is misleading to say that Galileo and his contemporaries discovered "the scientific method", it is certainly true that they developed methods and techniques which were to be immensely fruitful. As I have been emphasizing, the development of a method will typically itself constitute or be a consequence of a scientific discovery. For instance, Galileo's discovery of isochronism, the fact that the period of oscillation of a pendulum does not depend on the length of its swing, itself allowed the development of the pendulum clock by Huygens and Hooke. Similarly, Hooke's work on elasticity allowed the development of the spiral hairspring for the regulation of watches. Such cases show the leading scientists of the day combining basic research with the development of technology based upon it. In turn, this technology finds one of its applications in the accurate measurement of time for scientific purposes, furthering the development of mechanics and astronomy. The discovery of a fruitful method may lead not only to the acquisition of knowledge but also to the development of further fruitful methods.

Paradigms and progress

At various points in this book I have suggested that, in principle, scientific inference is holistic. We should take all our evidence into account, and we should seek that set of laws and explanations which overall is the best explanation of all this evidence. But in fact we never do exactly this. Typically a scientist considers only a restricted range of evidence and takes other theories for granted in using them to test a hypothesis concerning that hypothesis. Furthermore, a scientist's judgements as to what a good explanation in a certain field looks like will be guided by the history of successful research in that field. This will provide models of the various virtues of a theory. The leading theories will be exemplars of the general look and feel of a good hypothesis, including the "metaphysical principles" it should adopt (e.g. whether a field theory or a particle theory, whether action at a distance or contiguous cause and effect, whether an evolutionary or causal explanation). What counts as simplicity will be determined by those leading theories, while the success of a new theory will be in part determined by its ability to integrate with them. And they will show where interesting areas of research lie, which in turn will define fruitfulness for a theory in this field.

These sorts of consideration lead us to Thomas Kuhn's notion of a *paradigm*. Kuhn's view of the history of (a mature) science is that it consists of two kinds of episode: normal science and revolutionary science. During periods of normal science research is governed by certain standards and exemplars – the paradigm. Revolutions mark the transition between one period of normal science and another; they occur when a paradigm is overthrown and replaced. Revolutionary moments in the history of science include the triumph of the Copernican (heliocentric) model of the solar system over the Ptolemaic (geocentric) universe, the development of Newtonian mechanics, the discovery of oxygen and the end of phlogiston, the acceptance of the Darwinian theory of natural selection, and the replacement of classical physics by relativity theory and quantum mechanics.

It is because the idea of "guiding research" includes many factors that so many things have been called paradigms or aspects of them (which has led to disagreement about what Kuhn meant or should have meant by "paradigm"). The non-Kuhnian sense of "paradigm" is an exemplar, or standard, or perfect case. A person might be called a paradigm of virtue if their behaviour is to be emulated; Plutarch's lives were regarded for centuries as the paradigm of biographical writing and, as such, their form was taken as a template and their style copied by many writers; art students who copy the drawings of Raphael or musicians who learn to harmonize in the manner of a Bach chorale take the works of past masters as paradigms. This usual sense of paradigm is at the root of Kuhn's extended sense. (Kuhn later used "exemplar" to name the standard sense of "paradigm", and "disciplinary matrix" for the extended sense.) The works of great scientists are taken as paradigms of the way that science in its various branches ought to be done. Kuhn's examples include Newton's *Principia mathematica*, Lyell's *Geology*, Lavoisier's *Chemistry*, and Ptolemy's *Almagest*. A great text such as these will achieve various things: it will lay down the basic laws and principles of that science (e.g. Newton's laws of motion and gravitation, as found in the *Principia*); it will present certain problems (e.g. accounting for Kepler's laws) and so set the direction of future research; it will give solutions to those problems in terms of the basic laws, and thereby present a paradigm of good science; and certain methods (e.g. the calculus) may be introduced, which successors will similarly make use of.

The notion of a scientific paradigm has a strong sociological content. Great texts (or distillations of them) will be used in the education of young scientists. Students are made to repeat great experiments; they will be trained in the use of standard techniques, methods, and instruments. Inevitably these are the techniques to which they will turn first as practicing scientists; experimental arrangements will often be variants on the great experiments they copied as students. Furthermore, among their peers such practices will be regarded as marks of a good scientist; hence they tend to be perpetuated. The notion of a paradigm is clearly of importance to the philosopher too (as Kuhn intended it to be), because the paradigm sets not only the agenda of a science but its standards too. It is by reference to the paradigm that theories are evaluated.

A hitherto successful paradigm may find itself generating an increasing number of anomalies. The techniques and standard solutions of the paradigm are unable to explain away recalcitrant phenomena. Scientists themselves may begin to feel that the source of the problem is not an inability to find an answer within the framework set by the paradigm, but may lie with the paradigm itself. Kuhn draws a political parallel with institutions which generate political problems for which those institutions are unable to find an agreed resolution. Under such circumstances, dissatisfaction with institutional failure may lead to its being replaced – a revolution. When a revolution occurs the old institution provides no rules about how the new one should arise. Rather, that is a matter for collective renegotiation, persuasion, or force to decide. Similarly, revolutions in science occur when a paradigm is overthrown and a new one takes its place. While a paradigm may provide standards for the evaluation of theories which compete within that paradigm, it provides no rules for the evaluation of candidate paradigms seeking to replace it.

If this were exactly the case, then it would be impossible to say that a paradigm is better than the one it replaced or than competing paradigms which were not adopted. Nor would it be possible to say of theories from one paradigm that they are better than theories from another. The impossibility of comparing such theories is described by saying that they are *incommensurable*. That theories from different paradigms are incommensurable is the reason why Kuhn's picture of science is often called *relativist*, as standards of rationality are relative to a paradigm, not absolute.

In fact Kuhn thought that there are two sources of incommensurability. The first is, as we have seen, the fact that standards of theory comparison and evaluation are paradigm relative. The other is the claim that the meanings of the terms used in the competing theories or paradigms are not the same. So that, while two theories may seem to contradict one another, this is in fact a matter of equivocation. (This form of incommensurability was also advocated by Feyerabend.) Why should there be incommensurability owing to what is called *meaning variance*? Recall that in Chapter 4 we saw how positivists made a sharp distinction between theoretical terms and observational terms. A theoretical term gets its meaning from the theory in which it plays a part. Thus if we have two distinct theories then their terms will differ in meaning, even if the words are the same. Kuhn gives the example of the word "mass", which he says means something different when used by Newton from what it means when used by Einstein. According to Newton mass is conserved, while in Einstein's theory mass is not conserved but can be converted into and from energy.[67] Thus the two physicists must have meant different things by the term "mass". A variance of meaning between theoretical terms need not on its own be a problem. According to the positivists, what allows theories to compete is the fact that they have observational consequences (which can be stated without using theoretical concepts). It is in their observational consequences that theories can agree or disagree. So one might think that what allows relativity and classical mechanics to compete are their differing predictions about what would happen to a light ray in the vicinity of a massive object – and evidence relevant to the choice between the theories could be gained by making the appropriate observations (as Eddington did in 1919).

However, the problem arises when we consider the thesis of meaning dependence of observation on theory, which Feyerabend and Kuhn accepted. In their view there is no theory-independent observation language that can be used to express the observational consequences of competing theories in a way that is neutral between the theories. It now looks as if each theorist is caught within a web of theory, and that theorists in different webs are unable to find any neutral means of communication. If they are unable to communicate, then they are unable to disagree. This problem would be especially acute if the theory on which the observation depends is the same as the theory that the observation is intended to test. But this will not always, or even normally, be the case. So it seems as if there might be some language independent of the competing theories. A and B might be distinct theories, which have observational consequences O_A and O_B. O_A and O_B will be theory dependent, but they may be dependent on the same theory C, which is independent of both A and B. In such a case we will have a neutral language in which to compare some consequences of A and B. Nonetheless, such a comparison will depend on the availability of such a theory C. If A and B are low-level theories operating within the same paradigm, then there may be independent theories that can do this work. This might be the case if the competing theories are a bacterial theory and a viral theory of the nature of a certain disease. If, by contrast, the dispute is between the basic theories of competing paradigms, then there may be no such C. For instance, if we are comparing classical and relativistic physics, it will be difficult to find a relevant theory that is not related to one or the other.

The main objection to this line of thought starts from the fact that statements do not have to be couched in terms which mean precisely the same in order to contradict one another. "The largest sea creature is a fish" and "The baleen whale is not a fish" contradict one another, even though "the largest sea creature" and "the baleen whale" do not mean the same – it is sufficient that they refer to the same species. Nonetheless, Kuhn argues that not only does the word "mass" mean something different to Einstein and Newton, but also that the word when used in one theory refers (if it refers at all) to something different from what is referred to by the same word when used in the other theory. "Newtonian mass" and "Einsteinian mass" are not different ways of referring to the same stuff, but refer to different things altogether. Why should Kuhn have thought this? Two related views underpin Kuhn's thesis. The first is that a theoretical term has a sense which determines its reference; the second is that the sense of a theoretical term depends on the *whole* of the theory of which it is a part. Together these add up to the view, roughly speaking, that the theory amounts to one big description of its intended reference. Thus two distinct competing theories will constitute rival descriptions, which could not refer to the same thing. This is why Einsteinian mass and Newtonian mass cannot be the same thing, since part of the description of the latter is that it is always conserved, and part of the description of the former is that it is not always conserved.

This approach to the meaning of scientific terms, that it is a sense or description attached to terms which fixes their references or extensions, was one we examined in Chapter 3. I concluded that it was mistaken. Instead I proposed that what determines the reference is the *explanatory role* attributed to the natural, including theoretical, kind in question. This view also covers theoretical quantities such as mass. Furthermore, the explanatory role of a kind or quantity will not typically encompass the whole of a theory but only its core. So in the case of electrons, different theories (Bohr, Thomson, Rutherford) have attributed the same explanatory role to electrons – according to all of them electrons are whatever it is that is responsible for certain familiar electrical phenomena (e.g. electrostatics). On this view, it is wrong both to think that reference is determined by a description attached to an expression and to think that it is the whole of a theory which plays a part in determining the reference. Thus it is perfectly possible for competing theories to refer to the same entities or quantities, even if they differ and so contradict one another in their descriptions of those things. For this reason, incommensurability owing to meaning variance is not a general problem. This is consistent with accepting that scientific concepts can undergo change (for instance the explanatory role associated with a certain theoretical expression may change), and perhaps Kuhn's example of "mass" is an instance of this. That example may gain plausibility by being drawn from a conceptually complex part of physics; one cannot think that in the debate between creationists and Darwinians the two sides simply do not understand what the other is saying. (And, in any case, it is plausible that the explanatory role attributed to mass is the same for both Newton and Einstein, i.e. that mass is the quantity responsible for the inertial properties of things.)

There nonetheless remains the question of incommensurability owing to variance in standards of theory comparison and evaluation. If different paradigms and their contents really were incommensurable, it would certainly go against what scientists conceive of themselves as doing. Scientists who came to adopt

Einstein's relativistic mechanics or quantum theory thought of themselves as having reason to do so, although that may say more about their psychology than about the objective rationality of their choice. Kuhn later came to accept that there are five values to which scientists in all paradigms adhere: accuracy; consistency with accepted theories; broad scope; simplicity; and fruitfulness. However, this does not free us entirely from the problem of incommensurability; it still depends on the paradigm to fix what counts as fruitfulness, simplicity, accuracy, and so on; different paradigms may also weigh these desiderata differently. Take accuracy, for example. Accuracy is a matter of a theory fitting the data. Data are what we get from observation. But, according to Kuhn, what we observe is itself dependent on a paradigm. We encounter again the theory dependence of observation. Or, take simplicity. Someone might regard the mechanics of the special theory of relativity as comprising a set of elegantly simple equations; alternatively, one might regard the retreat from Euclidean geometry as giving up on simplicity.

One way in which theories can be incommensurable is by having different subject matters. The theories of continental drift and quantum electrodynamics explain different phenomena and so cannot be compared. Some more extreme followers of Kuhn, as well as some of his critics, have taken him as saying that, when a revolution occurs and a new paradigm ousts the old, the world has literally changed too; so that Priestley's phlogiston theory and Lavoisier's oxygen theory are literally about different things, and for this reason incommensurable. This interpretation makes Kuhn some sort of idealist – someone who denies that there is a world that exists independently of the way we believe it to exist. On a more reasonable interpretation, what Kuhn is pointing out is that the revolutionary nature of a paradigm shift makes it seem as if scientists operating in different paradigms are talking about different worlds. A new paradigm will make a scientist sensitive to new phenomena, furnish him with new instruments, and enable him to understand the world with new explanations, as well as give him a new vocabulary with which to describe them. And so Kuhn says of historians of science that they may be "*tempted* to exclaim that when paradigms change, the world itself changes with them".[68] And indeed one might be so tempted, especially if one takes "the world changing" in the colloquial sense of having quite a different outlook (or "world view").

Nonetheless, incommensurability among theories about the same phenomena does seem to have negative consequences for rational adherence to theories. On the one hand, theory comparison is licensed only within a paradigm, so there is no legitimate preference over theories from other paradigms. On the other hand, a preference among theories within a paradigm would be justified only if the theoretical assumptions of the paradigm are close to the truth and its standards of appraisal are truth-friendly. But we cannot stand outside the paradigm in order to assess these things (if we could, we could reasonably compare paradigms).

I think we can escape such anti-rationalist conclusions. First, incommensurability, whether true or not, does not entail that we have no scientific knowledge or justification. If the paradigm supplies reliable methods, then scientific knowledge is possible, even if we are not in a position to justify or evaluate the paradigm. If one takes a similarly externalist view of justification, a paradigm, if good enough, could provide justification for adopting a hypothesis, even if one had no way of saying how good the paradigm itself is. Incommensurability does lead to irrationalism if one thinks one must be able to justify one's justification, and so on. But I think that is a mistake, and the fact that justification may stop at the paradigm does not undermine that justification.

Secondly, we can accommodate many of the positive things that Kuhn says within the framework of Inference to the Best Explanation. We can do this without leading to the same conclusions concerning incommensurability. We have seen that Kuhn grants that his five values are values shared by different paradigms. They are all included in Newton-Smith's list of good-making features of a theory. I showed that all those features can be understood as aspects of inference to the best explanation. So it seems reasonable to conclude that all paradigms adhere to the principle of Inference to the Best Explanation. Which is what we would expect if Inference to the Best Explanation is the fundamental form of scientific inference. In Chapter 5, I discussed, in the context of the Duhem–Quine thesis, the fact that, although Inference to the Best Explanation is in principle holistic, it is typically applied in a local manner. This means that in a certain field a sequence of theoretical advances will be made with the principle of least disruption in mind. With each advance the theorist will intend to maintain as much of the existing theoretical structure while maximizing the quantity of new data explained and hence minimizing the number of new anomalies. However, the fact that each theoretical advance follows this rational procedure does not guarantee that at the end of the whole sequence we have the best available explanation of all the data in question.

For instance, it might be that had we made a different theoretical choice earlier on, that choice would have itself initially introduced many more anomalies and decreased the extent of the data satisfactorily explained. But it might have been that subsequent advances along this route would eventually have had more success than the later advances made in the path we actually did take.[69] If a theorist realizes that this is the position they are in, then it would make sense for them to take several steps back, to the point at which the path they took diverges from the one they might have taken. As the French biologist Jean Rostand put it, nothing leads the scientist so astray as a premature truth.

What is happening is that the scientists, in the face of increasing anomalies, are forced not to look for locally optimal explanations, but to take a more holistic view instead. In so doing, it may then appear that a radical rearrangement of the theoretical structure is required in order for it to be the best explanation available. This, I suggest, is what is happening in what Kuhn describes as a scientific revolution.

My reinterpretation of Kuhn has the advantage over the original that it is not beset by the problems of incommensurability. For the theorists are in a position to see and argue over whether the revolutionary rearrangement of the overall best explanation really is a good explanation. Nonetheless, there remain affinities between this view and the paradigm/incommensurability story. For scientists still operating at the local level, the revolution may not look like a theoretical advance. The revolution does not look like a stepwise improvement in the same direction, which is the sort of advance one would normally expect. Indeed, it may make recent step-wise, local advances look redundant. So there may be a strong psychological sense of revolution and, indeed, a psychological sense of incommensurability for some scientists – they just do not see why this is a sensible move to make. (Perhaps Priestley was in such a state of mind.) If a radical global rearrangement of the theoretical structure is a scientific revolution, step-wise progress corresponds to Kuhn's normal science. The theoretical core from which the step-wise progress starts will then be the paradigm. At a revolutionary moment, finding and assessing the appropriate new paradigm will be very difficult. Step-wise advances are easier to evaluate since nothing is being given up – almost all existing science will be accepted. But a revolutionary change will involve making a change at some point further back down the sequence of step-wise advances. It may not be clear how far back to go. The holistic assessment of a revolutionary theory as being the best explanation may be difficult to achieve. The quantity of relevant data will be much greater than is relevant to a step-wise advance. If the new theory can itself explain the general success of its predecessor, then assessing the data is that much easier, since the data successfully explained by the old theory can be commandeered as evidence for the new theory.

But the difficulties in finding, choosing, and justifying a new paradigm do not amount to impossibility. I have just mentioned the fact that new paradigms may be able to expropriate much of the data employed by the old paradigm. Presumably something like this happened with the shift from Newtonian to Einsteinian kinetics, since the latter can show that the behaviour of entities travelling at low relative velocities, which hitherto could be explained by Newtonian kinetics, can be explained by it too. If the two forms of physics were genuinely incommensurable, data from one would be irrelevant to the other. Secondly, in my view, the revolutionary shift does not typically require a complete theoretical reorganization. The new and the old will share some theoretical background. The new paradigm in geophysics, plate tectonics, shares with its predecessors common theoretical assumptions in terms of chemistry, physics, and geology.[70] Dalton's atomism did not require a rejection of previous chemistry – on the contrary, his hypothesis sought to explain the successes of that chemistry. And even if anti-atomists such as Sir Benjamin Brodie disagreed with Dalton and his followers, they were all able to share the same data. They were even able to see how the evidence impacted on the rival viewpoints. Brodie was able to see how the behaviour of isomers favoured the atomic hypothesis, even though he hoped to give a non-atomic account of the same phenomenon.[71]

These and other examples suggest that scientific revolutions do not involve incommensurability. What occurs is better explained by remarking that, guided by Inference to the Best Explanation, some theoretical proposals require a considerable rearrangement of the overall theoretical structure, as opposed to simply adding to it. But, as not everything is changed, there is enough in common to allow for a shared point of view between new and old or between adherents of rival proposals. This way of looking at things suggests that the dichotomy between normal science and revolutionary science is too stark. Some changes, such as Newton's, Einstein's, and Planck's, may have involved quite considerable theoretical upheaval. But others might require smaller, more localized rearrangements of theory – the discovery of X-rays, the inception of cryogenics, and Pauling's work on the chemical bond might be considered mid-level cases of constitutional development rather than revolutions.

Conclusion

I have argued that the optimistic view of scientific method is mistaken. There is no unique method that characterizes science, explains its successes, and is general in its application. Insofar as there are knowledge-producing methods in science, and there are many of them, they tend to be specific in their application and are discovered *a posteriori*.

Furthermore, if we think of methods in the sense of Chapter 7, as reliable or possibly reliable belief-forming processes, methods are supplemented by rules, principles, attitudes, preferences, and values which, although cannot be said to be methods, nonetheless guide and help direct belief formation. On the one hand, the litmus test is clearly a method which, as it happens, is reliable for its purposes. On the other hand, we have preferences, such as the preference for quantitative over purely qualitative theories. On its own this cannot constitute a method or belief-forming process, but it clearly contributes to belief formation. It is worth adding that even this preference is one the value of which is contingent and whose discovery is *a posteriori*. We could have inhabited a world in which the laws are qualitative; as it happens we do not, and this was something we had to discover.

I suspect that the optimistic view, or something like it, is part of what drives Hume's problem. If one thinks there is a unique scientific method, then if one asks what could justify use of this method, Hume's problem soon arises. But if one conceives of a large array of methods and principles, which are themselves justified or discovered by other methods, then Hume's problem, though not eliminated, is less pressing. I argued in Chapter 5 that the way in which Hume's problem is typically stated is misleading, in that it suggests that there is some single inference procedure, or induction, that deserves justifying. Correspondingly, in this chapter I have argued that there is no one scientific method. These two ideas contribute to one general thesis of the book, that it is a mistake to try and find models for explanation, induction, confirmation, method, and so on. Our approach to these things must, in principle if not in practice, be holistic. This follows from the fact that our concept of law is holistic.

The nearest we might get to a description, if not a model, of inductive inference is Inference to the Best Explanation. General proposals regarding what the principles of theory choice should be (Lakatos and Newton-Smith) can be seen as manifestations of Inference to the Best Explanation, while much of what Kuhn says actually happens can be interpreted in the same light. This claim does not detract from the previous paragraph. Inference to the Best Explanation is itself holistic. Any actual application of Inference to the Best Explanation will require experience and knowledge in the forming of appropriate possible explanations. They will also be required in deciding which is the best of the competing explanations.

Inference to the Best Explanation has to be holistic because our conception of laws is holistic. This is why we cannot expect to have a model of explanation and confirmation. Knowing what laws there are and what facts they explain go together. At any one time we have only seen some of the facts there are and thus only some of the instances of the laws of nature. Our best theories are an attempt to fit these facts into a pattern that best reflects what we expect a law-governed world to be like. But no one can ever look at all the facts we know in order to systematize them. Apart from being a herculean task, it is not always possible to know what it is we know. In practice, we must start from where others have left off. But we are not condemned merely to hope that they have got things right. Their success predisposes us to think they have, but failure to progress in the face of anomalies may force scientists of genius to take a broader, more inclusive view than has hitherto been possible. A revolution may ensue as science backtracks and reformulates some fundamental doctrines. But far from denying rationality and preventing progress, the ability of science both to criticize and thus to justify itself, is another reason for being confident that it is a rational, knowledge-generating enterprise.

Further reading

The classic texts discussed in this chapter are: Imre Lakatos' papers in *The methodology of scientific research programmes*, Paul Feyerabend's *Against method* and Thomas Kuhn's *The structure of scientific revolutions*. Bill Newton-Smith, in *The rationality of science,* gives very clear, critical discussions of Popper, Lakatos, Kuhn, and Feyerabend, as well as chapters on scientific method, incommensurability, and verisimilitude. A useful collection of readings is *Scientific revolutions,* edited by Ian Hacking.

[illegible]

Glossary

This glossary contains explanations of some of the philosophical and technical terms used in this book, or directions to where these are discussed in the text. A word of warning, however. Philosophers tend to disagree on the best definitions of many philosophical terms (just as they disagree on most other things). So the explanations I have given aim not to be definitive but to be helpful in understanding the terms as I have used them.

acceptance Some anti-realists (e.g. van Fraassen) argue that we should adopt an attitude to our best theories which is weaker than belief. Accepting a theory involves such things as working with it and reasoning with it, and having a research commitment to it.
***ad hoc* hypothesis** A hypothesis is called *ad hoc* if the reason for its existence is to explain away evidence which otherwise would falsify some favoured theory.
anomaly An anomaly is an observation, experimental result etc which conflicts with accepted theories.
anti-realism See Chapter 4.

a posteriori A proposition is known *a posteriori* if it is known, but not *a priori*. (See *a priori*.)
a priori This is an epistemological term, concerning kinds of knowledge. Roughly speaking, something is known *a priori* if it is known on the basis of reason alone and without any experience of the world. More accurately, a proposition is known *a priori* if it is known, without any more experience than is required to understand the proposition.
auxiliary hypothesis An hypothesis which is required in conjunction with a theory to generate observational consequences.
Bayesianism Bayesians take conditionalization using Bayes' theorem to be the basis of scientific inference. See Chapter 6.
Bayesian conditionalization Is the process of updating one's beliefs (in the subjective probabilities of hypotheses) in the light of new evidence in accordance with Bayes' theorem. If the latter tells us the probability of the hypothesis h given certain evidence e is k, then if we go on to obtain e, conditionalization means we should now attach probability k to h.
chance According to this book, chance is an objective property of things. An atomic nucleus can have a certain chance of decaying in a given period, and that is a law-governed property of that atom (not merely a statistical feature of a large number of such nuclei).
constructive empiricism See Chapter 4.
contingent A proposition is contingent if it is not necessary but is possible, i.e. it is one which might be true and might be false.
criterion If P is a criterion of Q, then P provides *a priori* good evidence for Q. See Chapter 1.
empirical Concerning what we can experience or observe.
empirically adequate A theory is empirically adequate if its observational consequences – what it says about things we could observe – are all true (note that empirical adequacy concerns all things we could observe, not just those we have observed).
empirically equivalent Two theories are empirically equivalent if they have the same observational consequences. And so, if one theory is empirically adequate, then the other will be too.
empiricism Empiricism is the view that our concepts all come from experience. According to empiricism we can have an idea of things which do not exist (such as "unicorn"), but only if these ideas are compounded out of elements which do correspond to

experience. Thus empiricists prefer minimalism about laws to full-blooded views involving necessitation, since necessitation cannot be experienced.

epistemic Concerning knowledge and justification.

epistemology That branch of philosophy which concerns itself with questions of knowledge, justified belief, scepticism, and so on.

essence The essence of kind is the set of properties a thing must have to be of that kind. If the essence is due to the meaning of the name of the kind then the essence is *nominal*. Some philosophers think that there are also *real* essences, owing to natural or metaphysical facts.

explanandum The fact that is to be explained.

explanans The fact that does the explaining.

extension The extension of a general term is the set of things which the term covers or is true of. So the extension of "mammal" is the set of all mammals.

externalism Externalist views of knowledge say that whether we know something depends in part on relations (with the thing known) that are external to the knower's awareness. Such relations might be causal relations or reliable methods, where the existence of the causation or the reliability of the method might themselves be unknown to the knower.

falsifiability A hypothesis is falsifiable if experimental evidence is conceivable which would require us to regard the hypothesis as false.

grue An adjective introduced by Nelson Goodman. Something is grue if and only if either it is observed before time *t* and is green, or is not observed before *t* and is blue (for some fixed time *t*, which in this book is midnight on 31 December 2000).

holistic In saying that, for example, explanation is holistic, it is being claimed that determining whether A is the explanation of B cannot be settled fully on a case by case basis, but only as part of determining a larger set of interrelated explanations.

Humean Relating to the philosophy of David Hume. In this book I have used "Humean" to refer to the idea that causation and lawhood are to be understood in terms of constant conjunction (regularities), and that induction proceeds by generalizing observed regularities.

incommensurable Two theories are incommensurable if they cannot be compared. This may be because there are no independent standards of evaluation, or because they cannot both be expressed in the same language.

induction See Introduction.

Inference to the Best Explanation (IBE) See Chapters 2 and 4.

intension A condition or description associated with an expression which determines its extension, for example the intension of "vixen" would be "female fox".

natural kind See Chapter 3.

necessary Following Leibniz, necessary propositions are those which are true in all possible worlds – they have to be true. To the extent that there are different conceptions of possibility there are different conceptions of necessity. For instance, we may be thinking of logically possible worlds, in which case a proposition that is true in all of them will be logically necessary. Or, instead, we may be thinking of nomically possible worlds – worlds that have the same laws of nature as ours, in which case a proposition which is true in all of these will be nomically necessary (though not logically necessary). (See *nomic necessity, contingent.*)

nomic Concerning laws of nature.

nomic necessity According to full-blooded, realist views of laws of nature, the universals in a law of nature are linked by a relation of nomic necessity. By virtue of nomic necessity, the presence of one universal explains the presence of the other. The existence or otherwise of a relation of nomic necessity is usually (but not by all) thought to be metaphysically contingent.

non-projectible A predicate is said to be non-projectible if it cannot be used in reasoning inductively from the observed to the unobserved, for example "grue".

ontological Concerning questions of existence.

paradigm See Chapter 8.

phenomena Roughly, that which is apparent and can be established by observation, although this can vary widely depending on context. Thus, for some philosophers "the phenomena" means the nature of our immediate experiences, while a scientist may regard the phrase as referring to the established features of a situation, especially regularities, even if these are heavily theoretical. Either way, the idea of the phenomena is in contrast to the idea of a reality behind the phenomena which explains them (which anti-realists of various sorts may deny exists).

phenomenological Concerning our experience of the world (as opposed to the world itself).

positivism Positivism is a version of empiricism which originates with Auguste Comte. One descendent, logical positivism, took science as its central concern. Positivism avoids metaphysics and what it sees as metaphysical speculation (e.g. on the existence of unobservable entities). It seeks an empirical foundation in experience, and takes the meaning of a sentence to be associated with the means by which it may be verified.

predicate A property or relation expression ("red", "between", "positively charged", etc.).

reliabilism Reliabilists about knowledge claim that knowledge is a true belief which rests on a reliable method. There is a corresponding reliabilist view of justification. Reliabilists are epistemic externalists.

theory laden An observation is theory laden, if, *either* the statement expressing the observation employs or presupposes certain theoretical concepts, or knowing the truth of the observation statement requires the truth of some theory. See Chapters 5 and 7.

truth-evaluable A sentence (or what it expresses) is truth-evaluable (or *truth-apt*) if it is either true or false. Some positivists regard theoretical statements as non-truth-evaluable.

truth-function One statement is a truth-function of others if the truth or falsity of the former is precisely determined by the truth and falsity of the latter. (The statement *p*, "It is raining and she is not coming", is a truth function of *q*, "It is raining", and *r*, "She is coming". *p* is true precisely when *q* is true and *r* is false.)

truth tropic A method of inference or belief formation is truth tropic if it tends to yield true conclusions or beliefs.

universal Universals are things like properties and relations which can be instantiated by more than one thing, such as "being yellow" and "being gravitationally attracted to".

verisimilitude Nearness to the truth. The concept of verisimilitude exists to catch the idea that, while two theories may both be false, one may still be better than the other by being closer to the way things are.

Notes

1. Gish & Bliss, *Summary of scientific evidence for creation.*
2. Gish & Bliss, *Summary of scientific evidence for creation.*
3. *McLean v. Arkansas Board of Education 529 F. Supp.* 1255 (E. D. Ark. 1982). Quoted in Feinberg (ed.), pp. 291–9.
4. Morris, *Studies in the Bible and science,* (Plaintiffs' exhibit 31), p. 114.
5. Gould, *The panda's thumb: more reflections in natural history* (Harmondsworth: Penguin, 1983).
6. Königsberg is a city in East Prussia (now part of Russia) where the River Pregel branches round an island, branches again, and then rejoins itself. There were seven bridges crossing these branches. The problem was whether it is possible to cross each of these bridges in one journey without crossing any bridge twice. Euler's solution to this problem marked the birth of topology.
7. Of course, someone might show that there is a flaw in a piece of mathematical reasoning. But that would not show that a proof can be overturned. Rather it shows that what we thought was a valid proof never was a valid proof. A valid proof is unassailable, whereas the very best scientific reasoning cannot render a theory immune from the logical possibility of falsifying evidence.

8. It must be admitted that scientists' (and philosophers') usage is a bit loose in this regard. We do talk about Kepler's laws and the ideal gas laws, and in doing so we must be talking about certain statements or theories – for there are no such laws, these theories being only close approximations to the truth.
9. Examples of this sort are to be found in the writings of Carl Hempel and Hans Reichenbach.
10. According to this view, the claim that all living dodos live in Australia is true; and it is also true that all living dodos live in Switzerland. Perhaps it is odd to regard such statements as true, but if we do not regard them as such we cannot count any of the empty generalizations about the artificial elements as being true – but some must be, namely those that correspond to laws.
11. Such accounts are provided by J. L. Mackie and David Lewis, whose views on counterfactuals are, despite superficial dissimilarities, closely related.
12. See Lewis, *Counterfactuals*, pp. 72–77. Lewis refers (p. 73) to an unpublished note by F. P. Ramsey.
13. This is the view favoured by David Lewis. See his *Counterfactuals,* p. 73.
14. This view is promoted by David Armstrong in his *What is a law of nature?*, pp. 52–9.
15. Points to this effect are made by Jeremy Butterfield in his review of Armstrong *What is a law of Nature?* in *Mind* **94**, 1985.
16. *Ontology* is that bit of philosophy which studies what sorts of thing exist.
17. It might have been that Jones' death just happened and has no explanation whatsoever. We will eventually have to take this into consideration.
18. Ruben, *Explaining explanation,* p. 187.
19. See Galison, *How experiments end*, p. 50.
20. In fact Eddington's observations were far from clear-cut. But his intention was clearly to confirm the theory of relativity by matching his observations with what is deducible from that theory.
21. We need to be careful, because this is not true as a general principle. In some cases evidence can provide some confirmation of inconsistent hypotheses. Evidence that a constant C lies between 10 and 20 may confirm both hypothesis 1, that $10 < C < 15$, and hypothesis 2, that $15 < C < 20$.
22. At least it would if it were a law that bats are all blind (which they are not); we need only assume that there is such a law to see how the example works.

23. "Morphology" and "anatomy" are often used synonymously, but sometimes morphology is distinguished as being the study of externally perceptible structure while anatomy concerns itself with internal structure.
24. Morphological differences within a species are called intraspecific variation. Obvious cases are differences between sexes and between old and young individuals. In the *Gilia tenuiflora* case the differences are between subspecies/varieties. John Dean, in his "Controversy over classification", discusses details of the *Gilia* cases, and John Dupré argues similar points in his "Natural kinds and biological taxa". Ernst Mayr discusses sibling species in depth in *Populations, species and evolution*, Ch. 3.
25. One piece of European legislation classifies carrots as fruit because the Portuguese make a jam out of carrots. This reflects a commercial necessity. Can we say this is unnatural?
26. I have changed some aspects of Kripke's examples.
27. Jadeite is a pyroxene with the formula $NaAlSi_2O_6$; nephrite is an amphibole with the formula $Ca_2(Mg, Fe)_5(Si_4, O_{11})_2(OH)_2$. Pyroxenes are metasilicates with single chains, while amphiboles have double chains; the two substances therefore have different cleavage angles. The two forms of jade are visually slightly different, the colour of jadeite being more intense than that of nephrite. But one need not regard the colour difference as sufficient to show that they are different substances, as many gemstones vary in appearance from one another but are regarded as belonging to the same kind, their colour differences being attributable to impurities.
28. It may be that of these only the concept of species is particularly important.
29. By organic chemists especially. One of the four branches of organic chemistry, the chemistry of aromatic compounds, takes its name from an olfactory classification.
30. Comte, *The positive philosophy of Auguste Comte*, p. 243.
31. See Brock, *The Fontana history of chemistry*, pp. 167–8.
32. A conditional is a statement of the form "if A then B" and a conjunction is a statement of the form "A and B". Both of these are examples of truth functions, i.e. propositions the truth of which depends in a systematic way on the truth of their constituents.
33. See Livingston, *Particle physics the high energy frontier*, pp. 65–8. For a longer, but interesting description of this experiment see Trigg, *Landmark experiments in twentieth century physics*, pp. 191–210.
34. Cartwright, *How the laws of physics lie*.

35. In the case of Kepler, because the perturbations due, among other things, to gravitational forces between planets are not described, and in Galileo's case because the rate of acceleration varies with distance from the surface of the Earth.
36. See Creary, "Causal explanation and the reality of natural component forces". Discussed in Cartwright, *How the laws of physics lie*, pp. 62–7.
37. Of these, scientists tend already to regard the electrostatic, magnetic, and nuclear weak forces as being aspects of one force.
38. Cartwright, *How the laws of physics lie*, p. 67.
39. Putnam, *Philosophical papers,* vol. i, *Mathematics, matter and method*, p. 73.
40. Van Fraassen, *The scientific image*, p. 71.
41. See Lipton, *Inference to the best explanation,* pp. 176–7.
42. A. Einstein & W. J. de Haas. Experimeteller Nachweis der Ampèrischen Molekularströme. *Verhandlung der deutschen physikalischen Gesellschaft Berichte* **13**, p. 17, 1915. Quoted in Galison, *How experiments end,* p. 51.
43. Lipton, *Inference to the best explanation*, pp. 122–32.
44. The word "mention" is ambiguous. Here I mean it in a strong sense where "mentioning something" implies making some categorical assertion about that thing either directly or indirectly via an assertion about a class of things that includes it. So "it will rain tomorrow" and "it will rain for the next week" both mention tomorrow, but "it either rained yesterday, or will rain tomorrow" does not, since the reference to "tomorrow" does not involve a categorical assertion about it.
45. Kant thought that some interesting synthetic (i.e. non-analytic, non-tautological) truths could be established by *a priori* reasoning. A Kantian may hope that the uniformity premise would have some such justification. But it is not clear how one might show this, and in any case, as I argue below, it is false.
46. Quine, "Two dogmas of empiricism" in Quine, *From a logical point of view*, 1953, p. 41.
47. Strawson, *Introduction to logical theory*, p. 257.
48. Peto, R. Aspirin after myocardial infarction" [editorial]. Lancet **i** (8179), pp. 1172–3, 1980.
49. See Williamson, "The broadness of the mental".
50. Bode's law is not a perfect example because it is not a perfect regularity. But it might have been. See pp. 29–30.
51. I say "rests on" rather than "brought about by" because one might have a belief brought about in the first place by a bad method but subsequently acquire good reasons for believing it.

52. See Papineau, *Reality and representation* pp. 137–8, and Peacocke, *Thoughts: an essay on content*, p. 140.
53. See, for instance, Earman, "Underdetermination, realism and reason", *Midwest studies in philosophy,* **18**, p. 37, 1993. Quoted in Williamson, Knowledge as evidence. *Mind* 1987. Williamson promotes the view that knowledge is at least as fundamental a notion as justification.
54. For this argument, see Miller, "Popper's qualitative theory of verisimilitude".
55. Lakatos, *The methodology of scientific research programmes* p. 50.
56. Lakatos, *The methodology of scientific research programmes*, p. 50.
57. Pocock, *Clinical trials – a practical approach,* p. 7.
58. Pocock, *Clinical trials – a practical approach,* p. 65 (original italics).
59. Newton-Smith, *The rationality of science.*
60. See Lipton, *Inference to the best explanation*, pp. 138–50.
61. Feyerabend, *Against method*, p. 28.
62. Chalmers, *Science and its fabrication.*
63. Quoted in Hempel, *Philosophy of natural science,* Ch. 5.
64. Copernicus, *De revolutionibus,* Lib. I, Cap 9. Quoted in Koestler, *The sleepwalkers,* p. 199.
65. Galileo, *Dialogue on the great world systems*, pp. 199–200.
66. Copernicus, *Letter against Werner*. Quoted in Koestler, *The sleepwalkers*, p. 203.
67. Kuhn, *The structure of scientific revolutions,* pp. 101–2.
68. Kuhn, *The structure of scientific revolutions*, p. 111 (my italics).
69. Of course, the right path might be unfruitful if taken too early. Atomism was not really successful until Dalton had the data available to make it a good explanation. It might be that a wrong theory is better, for a period, at generating fruitful ideas and data than is a right one.
70. This allowed plate tectonics to adopt theoretical developments (e.g. geosynclinal theory) that had been achieved before the inception of the new paradigm.
71. See p. 153 for explanation of isomers.

Bibliography

Achinstein, P. (ed). *Observation, experiment, and hypothesis in modern physical science* (Cambridge, MA: MIT Press, 1985).

Armstrong, D. *Belief, truth and knowledge* (Cambridge: Cambridge University Press, 1973).

Armstrong, D. *What is a law of nature?* (Cambridge: Cambridge University Press, 1983).

Armstrong, D. *Universals: an opinionated introduction* (Boulder, CO: Westview Press, 1989).

Barnes, B. & S. Shapin, *Natural order* (Beverly Hills, CA: Sage, 1979).

Bloor, D. *Knowledge and social imagery,* 2nd edn. (Chicago, IL: University of Chicago Press, 1991).

Bloor, D. & B. Barnes, Relativism, rationality and the sociology of knowledge. In Hollis & Lukes (eds), 1982.

Boyd, R. On the current status of scientific realism. In Leplin (ed.), 1984, and Boyd, Gaspar & Trout (eds), 1991.

Boyd, R. Observations, explanatory power and simplicity. In Boyd, Gaspar & Trout (eds), 1991, and Achinstein (ed.), 1985.

Boyd, R., P. Gaspar, J.D., Trout, (eds). *The philosophy of science* (Cambridge, MA: MIT Press, 1991).

Braithwaite, R. *Scientific explanation* (Cambridge: Cambridge University Press, 1953).

Brandon, R. & R. Burian (eds). *Genes, oganisms and populations* (Cambridge, MA: MIT Press, 1984).

Brock, W. *The Fontana history of chemistry* (London: Fontana, 1992).

Butler, R. J. (ed). *Analytical philosophy, first series* (Oxford: Blackwell, 1962).

Butterfield, J. 1985. Review of Armstrong *What is law of nature? Mind* **94**, pp. 164–6.

Cartwright, N. *How the laws of physics lie* (Oxford: Oxford University Press, 1983).

Chalmers, A. F. *Science and its fabrication* (Minneapolis: University of Minnesota Press, 1990).

Collins, H. *Changing order* (London: Sage, 1985).

Collins, H. & T. Pinch. *The golem: what everyone should know about science* (Cambridge: Cambridge University Press, 1993).

Comte, A. *The positive philosophy of Auguste Comte* (H. Martineau, trans.) (London: Chiswick, 1913).

Creary, L. Causal explanation and the reality of natural component forces. *Pacific Philosophical Quarterly* **62**, pp. 148–57, 1981.

Dean, J. Controversy over classification. In Barnes & Shapin 1979, pp. 211–30.

Dretske, F. Laws of nature. *Philosophy of Science* **44**, pp. 248–68, 1977.

Dupré, J. Natural kinds and biological taxa. *Philosophical Review* **90**, pp. 66–90, 1981.

Earman, J. *Bayes or bust* (Cambridge, MA: MIT Press, 1992).

Everitt, N. & A. Fisher. *Modern epistemology. A new introduction* (New York: McGraw Hill, 1995).

Feigl, H. & G. Maxwell, (eds). *Minnesota sudies in the philosophy of science vol. III.* (Minneapolis: University of Minnesota Press, 1962).

Feinberg, J. (ed.). *Reason and responsibility* (Belmont, CA: Wadsworth, 1996).

Feyerabend, P. Realism and instrumentalism, 1964. In Feyerabend, 1985.

Feyerabend, P. *Against method,* 3rd edn. (London: Verso, 1975/1993).

Feyerabend, P. *Philosophical papers,* vol 1. (Cambridge: Cambridge University Press, 1985).

Fine, A. The natural ontological attitude. In Leplin (ed.), 1984, and Papineau (ed.), 1996.

Fleck, L. *Genesis and development of a scientific fact* (Chicago: University of Chicago Press, 1979).

Friedman, M. Truth and confirmation. *Journal of Philosophy* **76**, pp. 361–82, 1979.

Galileo, G. *Dialogue on the great world systems* (Chicago, IL: University of Chicago Press, 1953).

Galison, P. *How experiments end* (Chicago: University of Chicago Press, 1987).

Gish, D. & R. Bliss, *Summary of scientific evidence for Creation.* CEI Cajon, CA: Institute for Creation Science, 1987).

Glymour, C. *Theory and evidence* (Chicago, IL: University of Chicago Press, 1981).

Glymour, C. Why I am not a Bayesian. In Papineau (ed), 1996.

Goodman, N. *Fact, fiction and forecast* (Atlantic Highlands, NJ: Athlone Press, 1954).

Gould, S. *The Panda's thumb: more reflections in natural history* (Harmondsworth: Penguin, 1983).

Grandy, R. (ed.). *Theories and observation in science* (Englewood-Cliffs, NJ: Prentice-Hall, 1973).

Hacking, I. Lakatos's philosophy of science. In Hacking (ed.), 1981.

Hacking, I. (ed). *Scientific revolutions* (Oxford: Oxford University Press, 1987).

Hacking, I. *Representing and intervening* (Cambridge: Cambridge University Press, 1983).

Hanson, N. *Patterns of discovery* (Cambridge: Cambridge University Press, 1972).

Harman, G. The inference to the best explanation. *Philosophical Review* **74**, pp. 88–95, 1965.

Harré, R. *The philosophies of science,* 2nd edn (Oxford: Oxford University Press, 1984).

Harré, R. & E.H. Madden, *Causal powers* (Oxford: Blackwell, 1975).

Hempel, C. Explanation in science and history. See Ruben (ed.), 1993.

Hempel, C. *Aspects of scientific explanation* (New York: Free Press, 1965).

Hempel, C. *Philosophy of natural science* (Englewood Cliffs, NJ: Prentice-Hall, 1966).

Hesse, M. *Revolutions and reconstructions in the philosophy of science* (Brighton: Harvester Press, 1980).

Hollis, J. & S. Lukes (eds). *Rationality and relativism* (Cambridge, MA: MIT Press, 1982).

Howson, C. & P. Urbach. *Scientific reasoning. A Bayesian approach*, 2nd edn (Chicago, IL: Open Court, 1993).

Hull, D. 1981. Units of Evolution. In Brandon and Burian (eds) 1984.

Hume, D. *A treatise of human nature* [1739] (Selby-Bigge, L.A. & P.H. Nidditch, eds) (Oxford: Oxford University Press, 1978, 2nd edn).

Hume, D. *Enquiry concerning human understanding* [1777] Selby-Bigge, L.A. & P.H. Nidditch, eds) (Oxford: Oxford University Press, 1975, 3rd edn.).

Kitcher, P. *Abusing science: the case against Creationism* (Cambridge, MA: MIT, 1982).

Koestler, A. *The sleepwalkers* (Harmondsworth: Penguin, 1959/ 1968).

Kornblith, H. (ed). *Naturalizing epistemology*, 2nd edn (Cambridge, MA: MIT Press, 1994).

Kripke, S. Identity and necessity. In Munitz (ed.), 1971.

Kripke, S. *Naming and necessity* (Cambridge, MA: Harvard University Press, 1980).

Kuhn, T. *The structure of scientific revolutions* (Chicago, IL: University of Chicago Press, 1962).

Kuhn, T. *The essential tension* (London: Chicago University Press, 1977).

Kuhn, T. Objectivity, value judgment and theory choice. In Kuhn, 1977.

Lakatos, I. *The methodology of scientific research programmes* (Cambridge: Cambridge University Press, 1977).

Lakatos, I. History of science and its rational reconstructions, 1970. In Hacking (ed.), 1981

Laudan, L. *Progress and its problems* (Berkeley, CA: University of California Press, 1977).

Laudan, L. A problem-solving approach to scientific progress. In Hacking (ed.), 1981.

Laudan, L. A confutation of convergent realism. In Leplin (ed.), 1984, and Boyd, Gaspar & Trout (eds), 1991.

Leplin, J. (ed.). *Scientific realism* (Berkeley, CA: University of California Press, 1984).

Lewis, D. *Counterfactuals* (Oxford: Blackwell, 1973).

Lipton, P. *Inference to the best explanation* (London: Routledge, 1991).

Livingston, M. S. *Particle physics: the high energy frontier* (New York: McGraw-Hill, 1968).

Locke, J. *An essay concerning human understanding* [1689–1690]. (A.D. Woozley, ed.) (London: Fontana, 1964, 5th edn).

Losee, J. *A historical introduction to the philosophy of science*, 2nd edn (Oxford: Oxford University Press, 1980).

Macdonald, G. & C. Wright (eds). *Fact, science and morality* (Oxford: Blackwell, 1987).

Mackie, J. Counterfactuals and causal laws. In Butler (ed.), 1962.

Mackie, J. *The cement of the universe* (Oxford: Oxford University Press, 1974).

Magee, B. *Popper* (London: Woburn Press, 1979).

Maxwell, G. The ontological status of theoretical entities. In Feigl & Maxwell (eds.), 1962.

Mayr, E. *Populations, species and evolution* (Cambridge, MA: Harvard University Press, 1975).

Mayr, E. *The growth of biological thought* (Cambridge, MA: Harvard University Press, 1982).

Mellor, D.H. *The matter of chance* (Cambridge: Cambridge University Press, 1971).

Mellor, D.H. Natural kinds. *British Journal of the Philosophy of Science* **28**, pp. 299–312, 1977. See Mellor 1991.

Mellor, D.H. The warrant of induction, 1987. In Mellor, 1991.

Mellor, D.H. *Matters of metaphysics* (Cambridge: Cambridge University Press, 1991).

Miller, D. Popper's qualitative theory of verisimilitude. *British Journal of the Philosophy of Science* **25**, pp. 178–88, 1974.

Morris, H. *Scientific creationism* (San Diego: Creation-Life, 1974).

Munitz, M. (ed.). *Identity and individuation* (New York: New York University Press, 1971).

Nagel, E. *The structure of science* (New York: Harcourt, Brace, 1961).

Newton-Smith, W. *The rationality of science* (London: Routledge & Kegan Paul, 1987).

Nola, R. (ed.). *Relativism and realism in science* (Dordrecht: Kluwer, 1988).

O'Hear, A. *Karl Popper* (London: Routledge & Kegan Paul, 1980).

O'Hear, A. *An introduction to the philosophy of science* (Oxford: Clarendon Press, 1989).

Papineau, D. *Theory and meaning* (Oxford: Clarendon Press, 1979).

Papineau, D. Realism and epistemology. *Mind* **94**, pp. 367–88, 1985.

Papineau, D. *Reality and representation* (Oxford: Blackwell, 1987).

Papineau, D. Laws and accidents. In Macdonald & Wright (eds), 1987.

Papineau, D. (ed.). *The philosophy of science* (Oxford: Oxford University Press, 1996).

Peacocke, C. *Thoughts, an essay on content* (Oxford: Blackwell, 1986).

Pocock, S. *Clinical trials – a practical approach* (Chichester: Wiley, 1983).

Popper, Sir K. *The logic of scientific discovery* (London: Hutchinson, 1959).

Popper, Sir K. *Conjectures and refutations* (London: Routledge and Kegan Paul, 1962).

Popper, Sir K. *Objective knowledge* (Oxford: Oxford University Press, 1972).

Putnam, H. The "Corroboration" of theories. In Hacking (ed.), 1981.

Putnam, H. *Philosophical papers:* vol. i, *Mathematics, matter and method*; vol. ii, *Mind, language and reality* (Cambridge: Cambridge University Press, 1975).

Putnam, H. The meaning of "meaning". In *Philosophical papers*, 1975, vol. ii.

Putnam, H. *Reason, truth and history* (Cambridge: Cambridge University Press, 1981).

Quine, W. van Orman. Two dogmas of empiricism. In Quine, 1953/1961.

Quine, W. van Orman. *From a logical point of view* (New York: Harper & Row, 1953/1961).

Quine, W. van Orman. Natural kinds. In *Ontological relativity and other essays* (New York: Columbia University Press, 1969). See Boyd et al., 1991, and Korblith, 1994.

Ramsey, F.P. Universals of law and universals of fact [1928]. In Ramsey 1990, pp. 140–4.

Ramsey, F.P. General propositions and causality [1929]. In Ramsey 1990, pp. 145–63.

Ramsey, F.P. *Philosophical papers* (D.H. Mellor, ed.) (Cambridge: Cambridge University Press, 1990).

Reichenbach, H. *Modern philosophy of science: selected essays* (M. Reichenbach, ed. and trans.) (London: Routledge & Kegan Paul, 1959).

Ruben, D.H. *Explaining explanation* (London: Routledge, 1990).

Ruben, D.H. (ed.). *Explanation* (Oxford: Oxford University Press, 1993).

Sankey, H. *The incommensurability thesis* (Aldershot: Avebury, 1994).

Schilpp, P.A. *The philosophy of Karl Popper*, 2 vols (La Salle: Open Court, 1974).

Skorupski, J. *John Stuart Mill* (London: Routledge, 1989).

Skyrms, B. *Choice and chance* (Belmont: Wadsworth, 1986).

Smart, J. *Philosophy and scientific realism* (New York: London: Humanities Press, Routledge, 1963).

Stevenson, L. & H. Byerly. *The many faces of science* (Oxford: Westview Press, 1995).

Stove, D.C. *Popper and after: four modern irrationalists* (Oxford: Pergamon, 1982).

Strawson, P.F. *Introduction to logical theory* (London: Methuen, 1952/1963).

Swinburne, R. *An introduction to confirmation theory* (London: Methuen, 1973).

Swinburne, R. (ed.). *The justification of induction* (London: Oxford University Press, 1974).

Tooley, M. The nature of laws. *Canadian Journal of Philosophy* **7**, pp. 667–98, 1977.

Trigg, G. *Landmark experiments in twentieth century physics* (New York: Crane, Russack, 1975).

van Fraassen, B. *The scientific image* (Oxford: Oxford University Press, 1980).

van Fraassen, B. To save the phenomena. In Boyd, Gaspar & Trout (eds), 1991, and Papineau (ed.), 1996.

van Fraassen, B. *Laws and symmetry* (Oxford: Oxford University Press, 1989).

von Mises, R. *Probability, statistics and truth* (London: Hodge, 1939).

von Wright, G.H. *The logical problem of induction,* 2nd edn. (New York: Macmillan, 1957).

von Wright, G.H. *Explanation and understanding* (Ithaca, NY: Cornell University Press, 1971).

Williamson, T. Knowledge as evidence. *Mind* **106**, pp. 717–741, 1997.

Williamson, T. The broadness of the mental. In *Philosophical perspectives*, vol. 12 (J. Tomberlin, ed.) (Oxford: Blackwell, 1998).

Woolgar, S. *Science: the very idea* (Chichester: Ellis Horwood, 1988).

Zemach, E. Putnam's theory on the reference of substance terms. *Journal of Philosophy* **73**, pp. 116–27, 1976.

Index